MICROBIOLOGICAL RISK ASSE

3

Hazard characterization for pathogens in food and water

GUIDELINES

WORLD HEALTH ORGANIZATION

FOOD AND AGRICULTURE ORGANIZATION OF THE UNITED NATIONS

2003

WHO Library Cataloguing-in-Publication Data

Hazard characterization for pathogens in food and water: Guidelines

(Microbiological risk assessment series No. 3)
1. Food microbiology 2. Water microbiology 3. Risk assessment - methods
4. Models, Statistical 5. Guidelines I. Joint FAO/WHO Secretariat on Risk
Assessment of Microbiological Hazards in Food II. Series

ISBN 92 4 156237 4 (WHO) (LC/NLM classification: QW 85)
ISBN 92 5 104940 8 (FAO)
ISSN 1726-5274

CONTENTS

ABBREVIATIONS USED IN THE TEXT

CAC	FAO/WHO Codex Alimentarius Commission
CFU	Colony forming unit
FAO	Food and Agriculture Organization of the United Nations
FDA	Food and Drug Administration [of the United States of America]
GDWQ	Guidelines for Drinking Water Quality
JEMRA	Joint FAO/WHO Expert Meetings on Microbiological Risk Assessment
MCMC	Markov chain Monte Carlo methods
MRA	Microbiological risk assessment
PCR	Polymerase chain reaction
USDA	United States Department of Agriculture
WHO	World Health Organization

A glossary of technical terms used in the text appears as Appendix B.

FOREWORD

The Members of the Food and Agriculture Organization of the United Nations (FAO) and of the World Health Organization (WHO) have expressed concern regarding the level of safety of food at both national and the international levels. Increasing foodborne disease incidence over the last decades seems, in many countries, to be related to an increase in disease caused by microorganisms in food. This concern has been voiced in meetings of the Governing Bodies of both Organizations and in the Codex Alimentarius Commission. It is not easy to decide whether the suggested increase is real or an artefact of changes in other areas, such as improved disease surveillance or better detection methods for microorganisms in foods. However, the important issue is whether new tools or revised and improved actions can contribute to our ability to lower the disease burden and provide safer food. Fortunately, new tools that can facilitate actions seem to be on their way.

Over the past decade, risk analysis – a process consisting of risk assessment, risk management and risk communication – has emerged as a structured model for improving our food control systems, with the objectives of producing safer food, reducing the numbers of foodborne illnesses and facilitating domestic and international trade in food. Furthermore, we are moving towards a more holistic approach to food safety, where the entire food chain needs to be considered in efforts to produce safer food.

As with any model, tools are needed for the implementation of the risk analysis paradigm. Risk assessment is the science-based component of risk analysis. Science today provides us with in-depth information on life in the world we live in. It has allowed us to accumulate a wealth of knowledge on microscopic organisms, their growth, survival and death, even their genetic make-up. It has given us an understanding of food production, processing and preservation, and the link between the microscopic and the macroscopic worlds and how we can benefit from as well as suffer from these microorganisms. Risk assessment provides us with a framework for organizing all this data and information and to better understand the interaction between microorganisms, foods and human illness. It provides us with the ability to estimate the risk to human health from specific microorganisms in foods and gives us a tool with which we can compare and evaluate different scenarios, as well as identify what types of data are necessary for estimating and optimizing mitigating interventions.

Microbiological risk assessment (MRA) can be considered a tool for use in the management of the risks posed by foodborne pathogens and in the elaboration of standards for food in international trade. However, undertaking an MRA, particularly quantitative MRA, is recognized as a resource-intensive task requiring a multidisciplinary approach. Yet, foodborne illness is among the most widespread public health problems, creating social and economic burdens as well as human suffering, making it a concern that all countries need to address. As risk assessment can also be used to justify the introduction of more stringent standards for imported foods, a knowledge of MRA is important for trade purposes, and there is a need to provide countries with the tools for understanding and, if possible, undertaking MRA. This need, combined with that of the Codex Alimentarius for risk-related scientific advice, led FAO and WHO to undertake a programme of activities on MRA at the international level.

The Food Quality and Standards Service, FAO, and the Food Safety Department, WHO, are the lead units responsible for this initiative. The two groups have worked together to develop the area of MRA at the international level for application at both national and international levels. This work has been greatly facilitated by the contribution of people from around the world with expertize in microbiology, mathematical modelling, epidemiology and food technology, to name but a few disciplines.

This Microbiological Risk Assessment Series provides a range of data and information to those who need to understand or undertake MRA. It comprises risk assessments of particular pathogen-commodity combinations, interpretative summaries of the risk assessments, guidelines for undertaking and using risk assessment, and reports addressing other pertinent aspects of MRA.

We hope that this series will provide a greater insight into MRA, how it is undertaken and how it can be used. We strongly believe that this is an area that should be developed in the international sphere, and from the present work already have clear indications that an international approach and early agreement in this area will strengthen the future potential for use of this tool in all parts of the world, as well as in international standard setting. We would welcome comments and feedback on any of the documents within this series so that we can endeavour to provide Member States, Codex Alimentarius and other users of this material with the information they need to use risk-based tools, with the ultimate objective of ensuring that safe food is available for all consumers.

Jean-Louis Jouve   Jørgen Schlundt

Food Quality and Standards Service   Food Safety Department
FAO   WHO

PREFACE

The process of developing hazard characterization guidelines was initiated at a workshop hosted by the WHO Collaborating Centre for Food Safety, National Institute for Public Health and the Environment (RIVM), Bilthoven, the Netherlands, 13–18 June 2000. The workshop participants were scientists actively involved in hazard characterization of food or waterborne pathogens in humans or animals. The document drafted during this workshop was reviewed at a series of expert meetings concerned with food and water safety, including:

- Joint FAO/WHO Expert Consultation on Risk Assessment of Microbiological Hazards in Foods, FAO Headquarters, Rome, Italy, 17–21 July 2000.

- WHO Meeting on Effective Approaches to Regulating Microbiological Drinking Water Quality, Glenelg, Adelaide, Australia, 14–18 May 2001.

A subsequent draft of the guidelines, incorporating these comments, was posted on the FAO and WHO Web sites, with a request for public comments. The draft guidelines were also circulated for peer review. The guidelines were then finalized, taking into consideration all comments received.

These guidelines have been written for an informed audience, and may be used in different contexts. In an international context, the guidelines will provide guidance for hazard characterizations conducted by the Ad hoc Joint Expert Meetings on Microbiological Risk Assessment and in the development of the WHO Guidelines for Drinking Water Quality (GDWQ). At the national level, they will provide guidance for hazard characterizations conducted for government and regulatory authorities.

ACKNOWLEDGEMENTS

The Food and Agriculture Organization of the United Nations and the World Health Organization would like to express their appreciation to all those who contributed to the preparation of this document through the provision of their time and expertize, data and other relevant information and by reviewing the document and providing comments.

Responsible technical units:

Food Quality and Standards Service   Food Safety Department

Food and Nutrition Division   World Health Organization

Food and Agriculture Organization of the
    United Nations

CONTRIBUTORS

<small>PARTICIPANTS AT THE BILTHOVEN WORKSHOP</small>

Robert L. Buchanan — Center for Food Safety and Applied Nutrition, United States Food and Drug Administration, United States of America.

James D. Campbell — Center for Vaccine Development, University of Maryland School of Medicine, United States of America.

Cynthia L. Chappel — Center for Infectious Diseases, University of Texas School of Public Health, United States of America.

Eric D. Ebel — United States Department of Agriculture, Food Safety and Inspection Service, United States of America.

Paul Gale — The National Centre for Environmental Toxicology, United Kingdom.

Charles N. Haas — Drexel University, School of Environmental Engineering and Policy, United States of America.

Arie H. Havelaar — WHO Collaborating Centre for Food Safety, National Institute of Public Health and the Environment, the Netherlands.

Jaap Jansen — Inspectorate for Health Protection, Commodities and Veterinary Public Health, the Netherlands.

Louise Kelly — Department of Risk Research, Veterinary Laboratories Agency, United Kingdom.

Anna Lammerding — Microbial Food Safety, Risk Assessment, Health Canada, Canada.

Roland Lindqvist — National Food Administration, Sweden.

Christine Moe — Department of Epidemiology, University of North Carolina, United States of America.

Roberta A. Morales — Research Triangle Institute, United States of America.

Mark Powell — United States Department of Agriculture, United States of America.

Stephen Schaub — United States Environmental Protection Agency, United States of America.

Wout Slob — National Institute of Public Health and the Environment, the Netherlands.

Mary Alice Smith — University of Georgia, United States of America.

Peter F.M. Teunis	National Institute of Public Health and the Environment, the Netherlands.
Sue Ferenc	International Life Sciences Institute (ILSI) Risk Science Institute, United States of America.
Jocelyne Rocourt	Laboratoire des *Listeria*, Institut Pasteur, France.
Johan Garssen	National Institute of Public Health and the Environment, the Netherlands.
Fumiko Kasuga	National Institute of Infectious Diseases, Japan.
Katsuhisa Takumi	National Institute of Public Health and the Environment, the Netherlands.
Andrea S. Vicari	North Carolina State University, United States of America.

REVIEWERS

Robert L. Buchanan	Center for Food Safety and Applied Nutrition, United States Food and Drug Administration, United States of America.
Charles N. Haas	Drexel University, United States of America.
Georg Kapperud	National Institute of Public Health, Norway.
Joan B. Rose	College of Marine Sciences, University of South Florida, United States of America.
Robert V. Tauxe	Foodborne and Diarrheal Diseases Branch, Centers for Disease Control and Prevention, United States of America.
Mark D. Sobsey	Division of Environmental Science and Engineering, University of North Carolina, United States of America.
Peter F.M. Teunis	National Institute of Public Health and the Environment, the Netherlands.
Desmond Till	Institute of Environmental Science & Research Ltd, Porirua, New Zealand.

TECHNICAL COORDINATOR

Arie H. Havelaar   WHO Collaborating Centre for Food Safety, National Institute of Public Health and the Environment, the Netherlands

THE JOINT FAO/WHO SECRETARIAT ON RISK ASSESSMENT OF MICROBIOLOGICAL HAZARDS IN FOODS

Jamie Bartram	WHO
Peter Karim Ben Embarek	WHO
Sarah Cahill	FAO
Maria de Lourdes Costarrica	FAO
Françoise Fontannaz	WHO
Allan Hogue	WHO (until July 2001)
Jean-Louis Jouve	FAO (from June 2001)
Hector Lupin	FAO
Jocelyne Rocourt	WHO (from January 2001)
Jørgen Schlundt	WHO
Hajime Toyofuku	WHO

TECHNICAL LANGUAGE EDITING

Thorgeir Lawrence, Iceland

1. INTRODUCTION

1.1 BACKGROUND

Microbiological risk assessment (MRA) is an emerging tool for the evaluation of the safety of food and water supplies. The Food and Agriculture Organization of the United Nations (FAO) and the World Health Organization (WHO) have a central role in developing and standardizing MRA at an international level, to inform risk management at both national and international levels. Elaboration of guidelines, such as these on Hazard Characterization for Pathogens in Food and Water, are important in achieving these tasks. The guidelines are primarily intended for a multidisciplinary audience, directly involved in developing and reviewing MRA documents at an international or national level. They will also be of use to risk managers who base their decisions on the risk assessment results, and need to be aware of the underlying principles and assumptions behind these assessments.

Hazard characterizations of microbiological pathogens in food and water are considered together in this document because the two cannot be effectively managed or understood in isolation from one another, despite historical differences in approaches. Water is both an ingredient of foodstuffs and an independent vehicle for human exposure to microbiological hazards through drinking, recreational activities or contact with aerosols. Pathways for human exposure to pathogens may involve both food and water, as illustrated in recent illness outbreaks resulting from wastewater re-use in the irrigation of fruits and vegetables. Reducing the public health impact of pathogens requires an understanding of the contributions of all primary routes of exposure. The use of a common approach for the characterization of microbiological hazards in food and water will foster this understanding, assist effective risk analysis and improve the protection of public health.

Ad hoc Joint FAO/WHO Expert Meetings on Microbiological Risk Assessment (JEMRA) conduct risk assessments of foodborne microbiological hazards at the international level. Risk management responsibilities for food in international trade are generally assigned to Codex Alimentarius Committees. JEMRA aims to provide a scientific basis for the relevant risk management deliberations of the Codex Alimentarius, whose purpose is to develop food standards, guidelines and related texts aimed at protecting consumer health and ensuring food fair trade practices. The JEMRA risk assessment reports also provide risk assessment information to FAO, WHO and Codex member countries.

International guidance on water quality and human health is provided by WHO through a series of guidelines. These include Guidelines for Drinking Water Quality (GDWQ), Guidelines for the Safe Use of Wastewater and Excreta in Agriculture and Aquaculture, and Guidelines for Safe Recreational Water Environments. These guidelines are based upon critical review of the best available scientific evidence and consensus, and derived by integrating information concerning adverse health effects with information concerning the effectiveness of safeguards under both normal and stressed conditions of operation (Fewtrell and Bartram, 2001). Their outputs are generalized standards for safety management that may include guidance both for good practice and for verifications (both analytical and inspection-based). These may then be adapted to take account of social, economic and environmental factors at local or national levels. Consequently, these guidelines provide the scientific basis for risk management activities of national and local policy-makers, assist in setting standards, and provide an international point of reference for the evaluation of water safety.

1.2 HAZARD CHARACTERIZATION IN CONTEXT

Risk analysis is a process comprising: risk assessment – the scientific evaluation of known or potential adverse health effects; risk management – evaluating, selecting and implementing policy alternatives; and risk communication – exchange of information amongst all interested parties. Although functional separation between these three components is important, there is increasing recognition of the necessity for interaction between them (FAO/WHO, 2002b; WHO, 2000b).

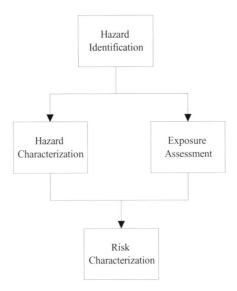

Figure 1. Components of a microbiological risk assessment

Risk assessment for microbiological hazards in foods is defined by the Codex Alimentarius Commission (CAC) as a scientifically based process consisting of four components (Figure 1): hazard identification, exposure assessment, hazard characterization, and risk characterization.

- **Hazard identification** is predominantly a qualitative process intended to identify microorganisms or microbial toxins of concern in food or water. It can include information on the hazard of concern as well as relevant related data, such as clinical and surveillance data.

- **Exposure assessment** should provide an estimate, with associated uncertainty, of the occurrence and level of the pathogen in a specified portion of food at the time of consumption, or in a specified volume of water using a production-to-consumption approach. While a mean value may be used, more accurate estimates will include an estimate of the distribution of exposures. This will typically include identification of the annual food and water consumption frequencies and weights or volumes for a given population or subpopulations(s), and should combine the information to estimate the population exposure to pathogens through a certain food or water commodity.

- **Hazard characterization** provides a description of the adverse health effects that may result from ingestion of a microorganism. When data are available, the hazard characterization should present quantitative information in terms of a dose-response relationship and the probability of adverse outcomes.

- **Risk characterization** is the integration of the three previous steps to obtain a risk estimate (i.e. an estimate of the likelihood and severity of the adverse health effects that would occur in a given population, with associated uncertainties).

The goal of a risk assessment may be to provide an estimate of the level of illness from a pathogen in a given population, but may also be limited to evaluation of one or several step(s) in a food production or processing system. When requesting a risk assessment, the risk manager should be specific with regard to the problem with which the risk manager needs to deal, the questions to be addressed by the risk assessment, and an indication of the measures the manager would consider or has available for the reduction of illness.

1.3 PURPOSE OF THE GUIDELINES

This present document is intended to provide a practical framework and a structured approach for the characterization of microbiological hazards, either in the context of a full microbiological risk assessment or as a stand-alone process. It is aimed at assisting governmental and research scientists to identify the points to be addressed, the methodology for incorporating data from different sources, and the methodology of dose-response modelling.

These guidelines are not a comprehensive source of information for hazard characterization. The expertize required spans several scientific disciplines, and a multidisciplinary team is essential to the endeavour. The issues involved are complex, in particular the methodology for dose-response modelling. Rather than specifying technical details, which are evolving at a rapid pace, reference is made to additional sources of information where appropriate. Modelling decisions may require consultation with experienced statisticians, mathematicians or experts in other scientific disciplines.

These guidelines are not intended to be prescriptive, nor do they identify pre-selected, compelling options. On some issues, an approach is advocated based on a consensus view of experts (e.g. the single-hit theory, bias-neutral approach to modelling – see Section 6.3.1) to provide guidance on the current science in hazard characterization. On other issues, the available options are compared and the decision on the approach appropriate to the situation is left to the analyst. In both of these situations, transparency requires that the approach and the supporting rationale be documented in the hazard characterization.

In the case of a hazard characterization conducted as a part of a complete risk assessment of pathogens in the food chain, the documentation developed by Codex Alimentarius on MRA provides the needed context. These present guidelines are intended to complement and provide additional detail to Codex documents, which provide more general guidance on MRA (e.g. CAC, 1999)

In the case of a hazard characterization conducted as a part of a complete risk assessment of pathogens in drinking water, GDWQ and the supporting background documents provide the necessary context.   In addition they provide information on the role of hazard characterization in the derivation of health-based targets for drinking-water quality.

1.4 SCOPE

These guidelines are limited to hazard characterization, considered either as a stand-alone process or as a component of MRA. They primarily address the effects on individual hosts of exposure to microbial pathogens. They do not consider the accumulation of individual risks over a population, nor risks of secondary transmission or the dynamic aspects thereof. These are elements of the risk characterization process.

These guidelines are limited to a consideration of the science of hazard characterization. No attempt is made to address risk management or risk communication issues, except to describe the interactions necessary to maximize the utility of the hazard characterization exercise (e.g. data collection, questions that need resolution, presentation of hazard characterization results). Issues related to establishing an appropriate level of protection are considered to be within the scope of risk management, and are not considered herein.

These guidelines are for the application of hazard characterization to microbial hazards in water and food. To date, most work has been aimed at pathogenic bacteria and viruses, and some parasitic protozoa. The principles outlined here, and in particular the descriptive methods, may also be applicable to other effects of single exposures to microorganisms or their toxins, including sequelae and chronic infections, such as by *Helicobacter pylori*. Effects of chronic exposure (e.g. to mycotoxins or algal toxins) may require another approach, which is more related to hazard characterization of toxic chemicals.

These guidelines focus on the adverse impact of a hazard as a result of ingestion of contaminated food and water. The adverse health effects that may occur as a result of exposure via other routes (e.g. through inhalation exposure) are not explicitly considered, but the basic principles outlined here might also be appropriate for characterizing alternative routes of exposure.

2. THE PROCESS OF HAZARD CHARACTERIZATION

2.1 CONTEXT

Hazard characterizations can be conducted as stand-alone processes or as component of risk assessment. A hazard characterization for a particular pathogen may serve as a common module or building block for risk assessments conducted for a variety of purposes and in an assortment of commodities. A hazard characterization developed in one country may serve the needs of risk managers in another country when combined with an exposure assessment specific to that country. A hazard characterization developed for water exposure may be adapted to a food exposure scenario by taking into consideration the food matrix effects. In general, hazard characterizations are fairly adaptable between risk assessments for the same pathogen. At the same time, exposure assessments are highly specific to the production, processing and consumption patterns within a country or region.

Hazard characterization, either as part of a risk assessment or as a stand-alone process, is iterative. Frequently the lessons learned lead to refinement of the initial question (or problem statement), leading in turn to further analysis. These guidelines for the characterization of hazards in food and water follow a structured, six-step approach, as outlined in Figure 2 and described in detail in subsequent chapters.

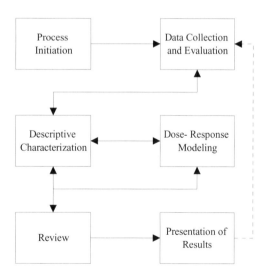

Figure 2. Process flow diagram for hazard characterization of pathogens

2.2 PRINCIPLES

Risk assessment and hazard characterization for microbial hazards should provide risk managers with a "best estimate" of the risk and the dose-response relationship, as free from bias as possible. Worst-case scenarios and deliberately conservative estimates reduce the utility of the risk estimate for cost–benefit or cost–effectiveness studies and decrease our ability to describe the uncertainty of the risk estimates. Uncertainty and variability should be tracked through the model to the extent possible, and included in the final estimate.

Independence and separation of hazard characterization from risk management are essential principles. Nevertheless, interaction between managers and assessors is also necessary to ensure the utility of the product to policy-makers and to ensure that the risk managers understand the principles and assumptions underlying the hazard characterization.

Transparency in hazard characterization requires full documentation of the process, including sources of data and their evaluation, and any assumptions made.

3. PROCESS INITIATION

Before beginning a risk assessment, the purpose and scope are established (i.e. problem formulation), but it may be useful to revisit these at the initiation of the hazard characterization step. The knowledge gained in previous work may, for example, indicate the need to refine the initial scope. That refinement often requires interaction with the risk managers to ensure that changes in the scope do not affect the utility of the final results.

The initiation of a hazard characterization requires a systematic planning stage to identify the context, purpose, scope and focus of the study to be carried out. Risk assessors should consider aspects of the pathogen, host, and food–water matrix (Figure 3).

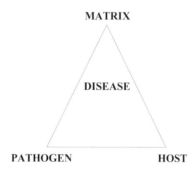

Figure 3. The epidemiology triangle (modified from Coleman and Marks, 1998)

Addressing the following list of questions may help structure or refine the problem under consideration:

- What are the characteristics of the pathogen that affect its ability to cause disease in the host (e.g. infectivity, pathogenicity, virulence)?

- What adverse health effects may be associated with exposure to the pathogen (from mild and self-limiting symptoms, to life-threatening conditions)?

- Who is susceptible to infection (individual/subpopulation/population)?

- What are the characteristics of the exposed population that may affect its susceptibility (age, immune status, concurrent illness, medical treatment, genetic background, pregnancy, nutritional status, social status, behavioural traits)?

- How frequently does infection give rise to clinical disease?

- What are the short- and long-term consequences (morbidity, mortality, sequelae, years of life lost, impairment of quality of life)?

- What are the most important routes of transmission?

- How does the response of the organism to environmental stress (heat, drying, pH, etc.) affect its ability to cause infection and illness?

- How does the matrix (food or water) affect the ability of the organism to cause infection and illness?

- Are multiple exposures independent or is some form of immune response likely?

In addition to the above parameters, it may be appropriate to include consideration of certain possible prevention or protection strategies, or both, such as immunization of a population against hepatitis A or typhoid fever.

These structuring questions need to be considered prior to the onset of hazard characterization. They are best formulated following a communication process between the assessors and the users of the hazard characterization results (risk managers). These questions will orient the collection, collation and evaluation of available information and data. They will also serve to identify data gaps and areas of uncertainty, and provide the risk manager with realistic expectations of the product of the hazard characterization. The answers provided should enhance knowledge of the pathogen and the disease, and identify areas where more research is needed.

A preliminary investigation phase may be necessary to define a risk model. In this phase, qualitative or quantitative evidence, or both, is structured to the overall framework of risk assessment and is used to "map" the risk model. Such studies may assess if sufficient data are available to answer the risk assessment question(s), and will result in a recommendation as to whether the assessment should be qualitative, quantitative, an analysis of the gaps in the data, or whether a different question might be answered with the available data.

4. DATA COLLECTION AND EVALUATION

Hazard characterizations are typically developed by compiling information from a variety of data sources, using a plethora of test protocols. Each of these data sources contributes in varying degrees to an understanding of the pathogen-host-matrix interactions that influence the potential public health risks attributable to different disease agents. An appreciation of the strengths and limitations of the various data sources is critical to selecting appropriate data for use, and to establishing the uncertainty associated with dose-response models that are developed from different data sets and test protocols.

Active data collection is required, because reliance on passive data submission or data in published form does not usually provide enough information in sufficient detail to construct dose-response models. Relevant data come preferably from peer-reviewed journals. Given the current lack of data for hazard characterization, it is also advisable to evaluate the availability of unpublished, high-quality data sources. Risk assessors should communicate with experimenters, epidemiologists, food or water safety regulatory persons, and others who may have useful data that could contribute to the analysis. An example of such is the outbreak information collected by the Japanese Ministry of Health and which was used for dose-response modelling of *Salmonella* (FAO/WHO, 2002a). When such data are used, the criteria and results of evaluation must be carefully documented. If using material published on the Internet, care should be taken to establish the provenance, validity and reliability of the data, and the original source, if known.

Understanding the characteristics of data sources is important to the selection and interpretation of data. Risk assessors often use data for a purpose other than that for which it was originally intended. Risk assessors and modellers need to know the means by which the data they use are collected, and the purpose of their collection. The properties of the available data will depend on the perspective of the researchers generating the data (e.g. experimenter versus epidemiologist). Therefore, knowledge of the source and original purpose of the available data sets is important in the development of dose-response models. The following sections attempt to capture in brief the strengths and limitations of each of several classes of data sources.

4.1 HUMAN STUDIES

4.1.1 Outbreak investigations

When there is a common-source outbreak of foodborne or waterborne disease of sufficient magnitude, an epidemiological investigation is generally undertaken to identify the cause of the problem, to limit its further spread, and to provide recommendations on how the problem can be prevented in the future. An outbreak of confirmed etiology that affects a clearly defined group can provide particularly complete information about the range of illness that a pathogen can cause, particular behaviour or other host characteristics that may increase or decrease the risk, and – if there is clinical follow up – the risk of sequelae. When the outbreak is traced to a food or water source that can be quantitatively cultured under circumstances that allow the original dose to be estimated, the actual dose-response can be measured. Even when that is not possible, dose-effect relations can often be observed that show variation in clinical response to changes in relative dose, and is part of the classic approach to an outbreak investigation. This may include looking for higher attack rates among persons who consumed more of the implicated vehicle, but may also include variation in symptom prevalence and complications. There are good public health reasons for gathering information on the amount of the implicated food or water consumed. An outbreak that is characterized by a low attack rate in a very large population may be an opportunity to define the host-response to very low doses of a pathogen, if the actual level of contamination in the food can be measured. In addition, data from outbreaks are the ultimate "anchor" for dose-response models and are an important way to validate risk assessments.

Strengths

An outbreak investigation can capture the diversity of host response to a single pathogenic strain. This can include the definition of the full clinical spectrum of illness and infection, if a cohort of exposed individuals can be examined and tested for evidence of infection and illness, independent of whether they were ill enough to seek medical care or diagnose themselves. It also includes definition of subgroups at higher risk, and the behaviour, or other host factors, that may increase or decrease that risk, given a specific exposure. Collecting information on underlying illness or pre-existing treatments is routine in many outbreak investigations.

Obtaining highly specific details of the food source and its preparation in the outbreak setting is often possible, because of the focus on a single food or meal, and may suggest specific correlates of risk that cannot be determined in the routine evaluation of a single case. Often, the observations made in outbreaks suggest further specific applied research to determine the behaviour of the pathogen in that specific matrix, handled in a specific way. For example, after a large outbreak of shigellosis was traced to chopped parsley, it was determined that *Shigella sonnei* grows abundantly on parsley left at room temperature if the parsley is chopped, but does not multiply if the parsley is intact. Such observations are obviously important to someone modelling the significance of low-level contamination of parsley.

Where samples of the implicated food or water vehicle can be quantitatively assayed for the pathogen, in circumstances that allow estimation of the original dose, an outbreak investigation has been a useful way to determine the actual clinical response to a defined dose in the general population.

Follow-up investigations of a (large) cohort of cases identified in an outbreak may allow identification and quantification of the frequency of sequelae, and the association of sequelae with specific strains or subtypes of a pathogen.

If preparations have been made in advance, the outbreak may offer a setting for the evaluation of methods to diagnose infection, assess exposure or treat the infection.

Limitations

The primary limitation is that the purpose and focus of outbreak investigations is to identify the source of the infection in order to prevent additional cases, rather than to collect a wide range of information. The case definitions and methods of the investigation are chosen for efficiency, and often do not include data that would be most useful in a hazard characterization, and may vary widely among different investigations. The primary goal of the investigation is to quickly identify the specific source(s) of infection, rather than to precisely quantify the magnitude of that risk. Key information that would allow data collected in an investigation to be useful for risk assessments is therefore often missing or incomplete. Estimates of dose or exposure in outbreaks may be inaccurate because:

- It was not possible to obtain representative samples of the contaminated food or water.

- If samples were obtained, they may have been held or handled in such a way after exposure occurred as to make meaningless the results of testing.

- Laboratories involved in outbreak testing are mainly concerned with presence/absence, and may not be set up to conduct enumeration testing.

- It is very difficult to detect and quantify viable organisms in the contaminated food or water (e.g. viable *Cryptosporidium* oocysts in water).

- Estimates of water or food consumption by infected individuals, and of the variability therein, are poor.

- There is inadequate knowledge concerning the health status of the exposed population, and the number of individuals who consumed food but did not become ill (a part of whom may have developed asymptomatic infection, whereas others were not infected at all).

- The size of the total exposed population is uncertain.

In such instances, use of outbreak data to develop dose-response models generally requires assumptions concerning the missing information. Fairly elaborate exposure models may be necessary to reconstruct exposure under the conditions of the outbreak. If microbiological risk assessors and epidemiologists work together to develop more comprehensive outbreak investigation protocols, this should promote the collection of more pertinent information. This might also help to identify detailed information that was obtained during the outbreak investigation but was not reported.

Even when all needed information is available, the use of such data may bias the hazard characterization if there are differences in the characteristics of pathogen strains associated with outbreaks versus sporadic cases. The potential for such bias may be evaluated by more detailed microbiological studies on the distribution of growth, survival and virulence characteristics in outbreak and endemic strains.

Estimates of attack rate may be an overestimate when they are based on signs and symptoms rather than laboratory-confirmed cases. Alternatively, in a case-control study conducted to identify a specific food or water exposure in a general population, the attack rate may be difficult to estimate, and may be underestimated, depending on the thoroughness of case finding.

The reported findings depend strongly on the case-definition used. Case definitions may be based on clinical symptoms, on laboratory data or a combination thereof. The most efficient approach could be to choose a clinical case definition, and validate it with a sample of cases that are confirmed by laboratory tests. This may include some non-specific illnesses among the cases. In investigations that are limited to culture-confirmed cases, or cases infected with a specific subtype of the pathogen, investigators may miss many of the milder or non-diagnosed illness occurrences, and thus underestimate the risk. The purpose of the outbreak investigation may lead the investigators to choices that are not necessarily the best for hazard characterization.

4.1.2 Surveillance and annual health statistics

Countries and several international organizations compile health statistics for infectious diseases, including those that are transmitted by foods and water. Such data are critical to adequately characterize microbial hazards. In addition, surveillance-based data have been used in conjunction with food survey data to estimate dose-response relations. It must be noted that, usually, analysis of such aggregated data requires many assumptions to be made, thus increasing uncertainty in results.

Strengths

Annual health statistics provide one means of both anchoring and validating dose-response models. The effectiveness of dose-response models is typically assessed by combining them with exposure estimates and determining if they approximate the annual disease statistics for the hazard.

Using annual disease statistics to develop dose-response models has the advantage that it implicitly considers the entire population and the wide variety of factors that can influence the biological response. Also, illness results from exposure to a variety of different strains. These data also allow for the relatively rapid initial estimation of the dose-response relationship. This approach is highly cost–effective since the data are generated and complied for other purposes. Available databases often have sufficient detail to allow consideration of special subpopulations.

Limitations

The primary limitation of the data is that they are highly dependent on the adequacy and sophistication of the surveillance system used to collect the information. Typically, public health surveillance for foodborne diseases depends on laboratory diagnosis. Thus it only captures those who were ill enough to seek care (and able to pay for it), and who provided samples for laboratory analysis. This can lead to a bias in hazard characterizations toward health consequences associated with the developed nations that have an extensive disease surveillance infrastructure. Within developed countries, the bias may be towards diseases with relatively high severity, that more frequently lead to medical diagnoses than mild, self-limiting diseases. International comparisons are difficult because a set of defined criteria for reporting is lacking at an international level. Another major limitation in the use of surveillance data is that it seldom includes accurate information on the attribution of disease to different food products, on the levels of disease agent in food and the number of individuals exposed. Use of such data to develop dose-response relations is also dependent on the adequacy of the exposure assessment, the identification of the portions of the population actually consuming the food or water, and the estimate of the segment of the population at increased risk.

4.1.3 Volunteer feeding studies

The most obvious means for acquiring information on dose-response relations for foodborne and waterborne pathogenic microorganisms is to expose humans to the disease agent under controlled conditions. There have been a limited number of pathogens for which feeding studies using volunteers have been carried out. Most have been in conjunction with vaccine trials.

Strengths

Using human volunteers is the most direct means of acquiring data that relates an exposure to a microbial hazard with an adverse response in human populations. If planned effectively, such studies can be conducted in conjunction with other clinical trials, such as the testing of vaccines. The results of the trials provide a direct means of observing the effects of the challenge dose on the integrated host defence response. The delivery matrix and the pathogen strain can be varied to evaluate food matrix and pathogen virulence effects.

Limitations

There are severe ethical and economic limitations associated with the use of human volunteers. These studies are generally conducted only with primarily healthy individuals between the ages of 18 and 50, and thus do not examine the segments of the human population typically most at risk. Pathogens that are life threatening or that cause disease only in high-risk subpopulations are not amenable to volunteer studies. Typically, the studies investigate a limited number of doses with a limited number of volunteers per dose. The dose ranges are generally high to ensure a response in a significant portion of the test population, i.e. the doses are generally not in the region of most interest to risk assessors.

The process of (self-)selection of volunteers may induce bias that can affect interpretation of findings. Feeding studies are not a practical means to address strain virulence variation. The choice of strain is therefore a critical variable in such studies. Most feeding studies use only rudimentary immunological testing prior to exposure. More extensive testing could be useful in developing susceptibility biomarkers.

Usually, feeding studies involve only a few strains, which are often laboratory domesticated or collection strains and may not represent wild-type strains. In addition, the conditions of propagation and preparation immediately before administration are not usually standardized or reported, though these may affect tolerance to acid, heat or drying, as well as altering virulence. For example, passage of *Vibrio cholerae* through the gastrointestinal tract induces a hyperinfectious state, which is perpetuated even after purging into natural aquatic reservoirs. This phenotype is expressed transiently, and lost after growth *in vitro* (Merrel et al., 2002). In many trials with enteric organisms, they are administered orally with a buffering substance, specifically to neutralize the effect of gastric acidity, which does not directly translate into what the dose response would be if ingested in food or water.

Considerations

In the development of experimental design, the following points need to be considered:

- How is dose measured (both units of measurement and the process used to measure a dose)?

- How do the units in which a dose is measured compare with the units of measurement for the pathogen in an environmental sample?

- Total units measured in a dose may not all be viable units or infectious units.

- Volunteers given replicate doses may not all receive the same amount of inoculum.

- How is the inoculum administered? Does the protocol involve simultaneous addition of agents that alter gastric acidity or promote the passage of microorganisms through the stomach without exposure to gastric acid?

- How do you know you dosed a naive volunteer (serum antibodies may have dropped to undetectable levels or the volunteer may have been previously infected with a similar pathogen that may not be detected by your serological test)?

- How is infection defined?

- What is the sensitivity and specificity of the assay used to determine infection?

- How is illness defined?

- When comparing the dose-response of two or more organisms, one must compare similar biological end-points, e.g. infection vs illness.

4.1.4 Biomarkers

Biomarkers are measurements of host characteristics that indicate exposure of a population to a hazard or the extent of adverse effect caused by the hazard. They are generally minimally invasive techniques that have been developed to assess the status of the host. The United States National Academy of Science has classified biomarkers into three classes, as follows:

- Biomarker of exposure – an exogenous substance or its metabolite, or the product of an interaction between a xenobiotic agent and some target molecule or cell, that is measured in a compartment within an organism.

- Biomarker of effect – a measurable biochemical, physiological or other alteration within an organism that, depending on magnitude, can be recognized as an established or potential health impairment or disease.

- Biomarker of susceptibility – an indicator of an inherent or acquired limitation of an organism's ability to respond to the challenge of exposure to a specific xenobiotic substance.

Even though this classification was developed against the background of risk assessment of toxic chemicals, these principles can be useful in interpreting data on pathogens.

Strengths

These techniques provide a means of acquiring biologically meaningful data while minimizing some of the limitations associated with various techniques involving human studies. Typically, biomarkers are measures that can be acquired with minimum invasiveness while simultaneously providing a quantitative measure of a response that has been linked to the disease state. As such, they have the potential to increase the number of replicates or doses that can be considered, or to provide a means by which objectivity can be improved, and increased precision and reproducibility of epidemiological or clinical data can be achieved. Biomarkers may also provide a means for understanding the underlying factors used in hazard characterization. A biomarker response may be observed after exposure to doses that do not necessarily cause illness (or infection). Biomarkers can be used either to identify susceptible populations or to evaluate the differential response in different population subgroups.

It should also be noted that the most useful biomarkers are linked to illness by a defined mechanism, that is, the biological response has a relationship to the disease process or clinical symptom. If a biomarker is known to correlate with illness or exposure, then this information may be useful in measuring dose-response relationships, even if the subjects do not develop clinical symptoms. Biomarkers such as these can be used to link animal studies with human studies for the purposes of dose-response modelling. This is potentially useful because animal models may not produce clinical symptoms similar to humans. In which case, a biomarker may serve as a surrogate end-point in the animal.

Limitations

Biomarkers are often indicators of infection, illness, severity, duration, etc. As such, there is a need to establish a correlation between the amplitude of the biomarker response and illness conditions. Biomarkers primarily provide information on the host status, unless protocols are specifically designed to assess the effects of different pathogen isolates or matrices.

The only currently available biomarkers for foodborne and waterborne pathogens are serological assays. The main limitation for such assays is that, in general, the humoral immune response to bacterial and parasitic infections is limited, transient and non-specific. For example, efforts to develop an immunological assay for *Escherichia coli* O157 infections have shown that a distinctive serological response to the O antigen is seen typically in the most severe cases, but is often absent in cases of culture-confirmed diarrhoea without blood. In contrast, serological assays are often quite good for viruses. Other biomarkers, such as counts of subsets of white blood cells or production of gaseous oxides of nitrogen are possible, but have not been tested extensively in human populations.

4.1.5 Intervention studies

Intervention studies are human trials where the impact of a hazard is evaluated by reducing exposure for a defined sample of a population. The incidence of disease or the frequency of a related biomarker is then compared to a control population to assess the magnitude of the response differential for the two levels of exposure.

Strengths

Intervention studies have the advantage of studying an actual population under conditions that are identical to or that closely approach those of the general population. In such a study, the range of host variability is accommodated. These studies are particularly useful in evaluating long-term exposures to levels of pathogens to which the consumer is likely to be subjected. Since intervention studies examine the diminution of an effect in the experimental group, the identified parameters would implicitly include the pathogen, host and food-matrix factors that influence the control groups. Potentially, one could manipulate the degree of exposure (dose) by manipulating the stringency of the intervention.

Limitations

Since exposure for the control group occurs as a result of normal exposure, the pathogen, host and food-matrix effects are not amenable to manipulation. Great care must be given to setting up appropriate controls, and in actively diagnosing biological responses of interest in both the test and control populations. It is often the case that intervention studies result in declines in response that are less than those predicted by the initial exposure. This is often due to the identification of a second source of exposure or an overestimation of the efficacy of the intervention. However, such data by itself is often of interest.

Considerations

Testable interventions – i.e. feasible in terms of technical, cultural and social issues – are "conservative" in that there are ethical boundaries. Thus they must be implemented within a defined population and, apart from being technically feasible, must be socially acceptable and compatible with the preferences and technical abilities of this population.

4.2 ANIMAL STUDIES

Animal studies are used to overcome many of the logistical and ethical limitations that are associated with human-volunteer feeding studies. There are a large variety of different animal models that are used extensively to understand the pathogen, host and matrix factors that affect characteristics of foodborne and waterborne disease, including the establishment of dose-response relations.

Strengths

The use of surrogate animals to characterize microbial hazards and establish dose-response relations provides a means for eliminating a number of the limitations of human-volunteer studies while still maintaining the use of intact animals to examine disease processes. A number of animal models are relatively inexpensive, thus increasing the potential for testing a variety of strains and increased numbers of replicates and doses. The animals are generally maintained under much more controlled conditions than human subjects. Immunodeficient animal strains and techniques for suppressing the immune system and other host defences are available and provide a means for characterizing the response in special subpopulations. Testing can be conducted directly on animal subpopulations such as neonates, aged or pregnant populations. Different food vehicles can be investigated readily.

Limitations

The major limitation is that the response in the animal model has to be correlated with that in humans. There is seldom a direct correlation between the response in humans and that in animals. Often, differences between the anatomy and physiology of humans and animal species lead to substantial differences in dose-response relations and the animal's response to disease. For a number of diseases, there is no good animal model. Several highly effective models (e.g. primates or pigs) can be expensive, and may be limited in the number of animals that can be used per dose group. Some animals used as surrogates are highly inbred and consequently lack genetic diversity. Likewise, they are healthy and usually of a specific age and weight range. As such, they generally do not reflect the general population of animals of that species, let alone the human population. Ethical considerations in many countries limit the range of biological end-points that can be studied.

Considerations

- When surrogate pathogens or surrogate animal models are used, the biological basis for the use of the surrogate must be clear.

- Using data obtained with animal models to predict health effects in humans could take advantage of the use of appropriate biomarkers.

- It is important to use pathogen strains that are identical or closely related to the strain of concern for humans, because, even within the same species and subspecies, different strains of pathogens may have different characteristics that cause variation in their abilities to enter and infect the host and cause illness.

4.3 *IN VITRO* STUDIES

In vitro studies involve the use of cell, tissue or organ cultures and related biological samples to characterize the effect of the pathogen on the host. They are of most use for qualitative investigations of pathogen virulence, but may also be used to evaluate in detail the effects of defined factors on the disease process.

Strengths

In vitro techniques can readily relate the characteristics of a biological response with specific virulence factors (genetic markers, surface characteristics and growth potential) under controlled conditions. This includes the use of different host cells or tissue cultures to represent different population groups, and manipulation of the environment under which the host cells or tissues are exposed to the pathogen, in order to characterize differences in dose-response relations between general and special populations. *In vitro* techniques can be used to investigate the relations between matrix effects and the expression of virulence markers. Large numbers of replicates and doses can be studied under highly controlled conditions.

These techniques can be used to readily compare multiple species and cell types to validate relationships between humans and surrogate animals. They are particularly useful as a means of providing information concerning the mechanistic basis for dose-response relations.

Limitations

The primary limitation is the indirect nature of information concerning dose-response relations. One cannot directly relate the effects observed with isolated cells and tissues to disease conditions that are observed within intact humans, such as the effect of integrated host defences. To compare with humans, there is need for a means to relate the quantitative relations observed in the *in vitro* system to those observed in the host. These types of studies are usually limited to providing details of factors affecting dose-response relations and to augmenting the hazard characterization, but are unlikely to be a direct means of establishing dose-response models useful for risk assessments. For many organisms, the specific virulence mechanisms and markers involved are unknown, and may vary between strains of the same species.

4.4 EXPERT ELICITATION

Expert elicitation is a formal approach to the acquisition and use of expert opinions, in the absence of or to augment available data.

Strengths

When there is a lack of the specific data needed to develop dose-response relations, but there are scientific experts with knowledge and experience pertinent to the elucidation of the information required, expert elicitation provides a means of acquiring and using this information so that consideration of dose-response relations can be initiated. This can involve the development of a distribution for a parameter in a model for which there is no, little or inconsistent numerical data, through the use of accepted processes that outline the lines of evidence or weight of evidence for generation of the opinion and use of the results. It is generally not expensive, particularly in relation to short-term needs.

Limitations

Results obtained are dependent on the methodology used, and are inherently subjective and thus open to debate. The results are also dependent on the experts selected and may have limited applicability for issues involving an emerging science.

4.5 DATA EVALUATION

Risk assessors must evaluate both the quality of the available sources of data for the purpose of the analysis, and the means of characterizing the uncertainty of all the data used. Formalized quality control of raw data and its subsequent treatment is desirable, but also highly dependent on availability and the use to which the data are applied. There is no formalized system for evaluation of data for hazard characterization. Few generalizations can be made, but the means by which data are collected and interpreted needs to be transparent. "Good" data are complete, relevant and valid: complete data are objective; relevant data are case-specific; and validation is context specific.

Complete data includes such things as the source of the data and the related study information, such as sample size, species studied and immune status. Characteristics of relevant data include age of data; region or country of origin; purpose of study; species of microorganism involved; sensitivity, specificity and precision of microbiological methods used; and data collection methods. Observations in a database should be "model free" – i.e. reported without interpretation by a particular model – to allow data to be used in ways that the original investigator might not have considered. This may require access to raw data, which may be difficult to achieve in practice. Using the Internet for such purposes should be encouraged, possibly by creating a Web site with data sets associated with published studies.

Valid data is that which agrees with others in terms of comparable methods and test development. In general, human data need less extrapolation and are preferred to animal data, which in turn are preferable to *in vitro* data. Data on the pathogen of concern are preferred to data on surrogate organisms, which should only be used on the basis of solid biological evidence, such as common virulence factors.

Currently, the recommended practice is to consider all available data as a potential source of information for hazard characterization. Data that can be eliminated from the risk assessment depends on the purpose and stage of the assessment. In the early stages of risk assessment, small data sets or those with qualitative values may be useful, whereas the later stages of risk assessment may include only those data that have been determined to have high quality standards. Excluding data from the analysis should be based on predefined criteria, and not based solely on statistical criteria. If the analysis is complicated by extreme heterogeneity or by outliers, it is advisable to stratify the data according to characteristics of the affected population, to microbial species, to matrix type or to any other suitable criterion. This practice should provide increased insight rather than information loss.

Sources of data are either peer-reviewed or non-peer-reviewed literature. Although peer-reviewed data are generally preferable for scientific studies, they also have some important drawbacks as inputs for dose-response modelling. First and foremost, they have limited availability. Also, important information may be missing concerning how dose and response data were obtained, as outlined here below. Data presentation in the peer-reviewed literature is usually in an aggregated form, not providing the level of detail necessary for uncertainty analysis. In older papers, the quality control of the measurement process may be poorly documented. For any of these reasons, the analyst might wish to add information from other sources. In that case, the quality of the data should be explicitly reviewed, preferably by independent experts.

An important aspect with regard to *dose information* is the performance characteristics of the analytical method. Ideally, a measurement reflects with a high degree of accuracy the true number of pathogens in the inoculum. Accuracy is defined as the absence of systematic error (trueness) and of random error (precision). Trueness of a microbiological method is defined by the recovery of target organisms, the inhibitory power against non-target organisms, and the differential characteristics of the method, as expressed in terms of sensitivity and specificity. Precision is related to the nature of the test (plating vs enrichment), the number of colonies counted or the number of positive subcultures, and the dispersion of the inoculum in the test sample (see Havelaar et al., 1993). It is also important to know the variation in ingested dose between individuals, related to the dispersion of the pathogens in the inoculum, but also in relation to different quantities of the inoculum being ingested. These characteristics are of particular relevance when using observational data on naturally occurring infections. A pathogen's infectivity can be affected by both the matrix and the previous history of the pathogen, and this should be taken into account.

With regard to *response information*, it is important to note whether the outcome was represented as a binary or a continuous outcome. Current dose-response models (see Chapter 6) are applicable to binary outcomes, and this requires that the investigator define the criteria for both positive and negative responses. The criteria used for this differentiation may vary between studies, but should explicitly be taken into account. Another relevant aspect is the characteristics of the exposed population (age, immunocompetence, previous exposure, etc.).

The aspects listed in this section are not primarily intended for differentiating "good" from "bad" data for hazard characterization, but rather to guide the subsequent analysis and the use of the dose-response information in a risk assessment model.

5. DESCRIPTIVE CHARACTERIZATION

Descriptive hazard characterization serves to structure and present the available information on the spectrum of human illness associated with a particular pathogen, and how this is influenced by the characteristics of the host, the pathogen and the matrix, as indicated in Figure 3. This is based on a qualitative or semi-quantitative analysis of the available evidence, and will take the different pathogenic mechanisms into account.

5.1 INFORMATION RELATED TO THE DISEASE PROCESS

When a hazard characterization is being undertaken, one of the initial activities will be to evaluate the weight of evidence for adverse health effects in humans in order to determine, or confirm, the ability of the pathogen to cause disease. The weight of evidence is assessed on the basis of causality inferences appropriately drawn from all of the available data. This entails examination of the quantity, quality and nature of the results available from clinical, experimental and epidemiological studies; analyses of pathogen characteristics; and information on the biological mechanisms involved. When extrapolating from animal or *in vitro* studies, awareness of the biological mechanisms involved is important with respect to assessment of relevance to humans.

Undertaking hazard characterization for waterborne and foodborne microbial pathogens, the biological aspects of the disease process should be considered. Each of these steps is composed of many biological events. Careful attention should be given in particular to the following general points:

- The process as a whole as well as each of the component steps will vary by the nature of the pathogen.

- Pathogens may be grouped in regard to one or more component steps, but this should be done cautiously and transparently.

- The probability of an event at each step may be dependent or independent of other steps.

- The sequence and timing of events are important.

For (toxico-)infectious pathogens, it is recommended to consider separately the factors related to infection and those related to illness as a consequence of infection (discussed later, in Chapter 6). While doing so, the following points should be considered:

- A definition of infection is not universally accepted.

- Infection is difficult to measure and depends on the sensitivity of diagnostic assay.

- Target cells or tissue may be specific (one cell type) or non-specific (many cell types), and local (non-invasive) or invasive or systemic, or a combination.

- The sequence of events and the time required for each may be important and may vary according to pathogen.

- Infection can be measured dichotomously (infection: yes or no), but some aspects can be measured quantitatively.

The information related to the disease should provide detailed – qualitative or quantitative, or a combination – insights into the disease process. In most cases, this would be based on the available clinical and epidemiological studies. Narrative statements are helpful to summarize the nature of and confidence in the evidence, based on limitations and strengths of the database. Each source of information has its advantages and limitations, but collectively they permit characterization of potential adverse health effects. The analysis should include evaluations of the statistical power of the studies, and appropriate control of possible bias, while identifying what is uncertain and the sources of uncertainty.

Characterization of the adverse human health effects should consider the whole spectrum of possible effects in response to the microbial hazard, including asymptomatic infections and clinical manifestations, whether acute, subacute or chronic (e.g. long-term sequelae), or intermittent (see Table 1). Where clinical manifestations are concerned, the description would include consideration of the diverse clinical forms, together with their severity, which

Table 1. Elements that might be included in characterization of adverse human health effects.
Clinical forms
Duration of illness
Severity (morbidity, mortality, sequelae)
Physiopathology
Epidemiological pattern
Secondary transmission
Quality of life
SOURCE: Adapted from ILSI, 2000.

may be variable among pathogens and among hosts infected with the same pathogen. Severity may be defined as the degree or extent of clinical disease produced by a microorganism, and may be expressed in a variety of ways, most of which include consideration of possible outcomes. For mild gastrointestinal symptoms, severity may be expressed as duration of the illness, or as the proportion of the population affected (morbidity). Where the gravity of the distress requires medical care or includes long-term illness, or both, severity may be expressed in terms of the costs to society, such as the proportion of workdays lost or cost of treatment. Some pathogens and the related clinical forms may be associated with a certain degree of mortality and therefore severity may be expressed as mortality rate. For pathogens that cause chronic illness (i.e. the disease leaves long-term sequelae) it may be desirable to include, in the characterization of the human health effects, considerations related to quality of life as it may be affected by the disease. Quality of life may be expressed in a variety of ways, depending on the nature of the illness. For instance, human life expectancy may decrease, chronic debilitation may occur, or quality of life may be affected by episodic bouts of disease. Increasingly, concepts such as Quality Adjusted Life Years or Disability Adjusted Life Years are being used to integrate and quantify the effects of different disease end-points on the health of individuals or populations (for examples, see WHO, 2000a; Havelaar et al., 2000).

In addition to a description of the human adverse health effects, information on the disease should include consideration of the epidemiological pattern and indicate whether the disease may be sporadic, endemic or epidemic. The frequency or incidence of the disease or its clinical forms, or both, should be addressed, together with their evolution with time and possible seasonal variations. The description should include consideration of the repartition of clinical forms according to specific groups at risk. Finally, the potential for, extent of or amount of transmission, including asymptomatic carriers, as well as secondary transmission, should also be characterized. Information collected on these aspects is important to guide the risk characterization phase of the risk assessment.

In all cases, and with particular regard to further modelling, it is important that the characterization includes a definition of possible end-points to be considered. Thought needs to be given to the appropriate criteria when defining "infection" of the host by the pathogenic agent, and the criteria of what constitutes a clinical "case". In addition, a definition of the severity scale should be provided, specifying the indicator chosen (e.g. disease end-point or consequences) and how it can be measured. The description should also include information on uncertainties and their sources.

To the extent possible, the characterization should incorporate information on the physiopathology of the disease, i.e. on the biological mechanisms involved. Depending on the information available, this would include consideration of elements such as:

- the entrance route(s) of a microorganism into a host;
- the effect of growth conditions on expression of virulence by and survival mechanisms of the microbe;
- the influence of the conditions of ingestion, including matrix effects;
- the influence of gastrointestinal status;
- the mechanisms involved in the penetration of the pathogen into tissues and cells;
- the status of the pathogen relative to non-specific cell-mediated (innate) immunity;
- the status of the pathogen relative to humoral defences;
- the effect of intercurrent illnesses and treatments, such as immunosuppressive or antimicrobial therapy;
- the potential for natural elimination; and
- the behaviour of the pathogen in a host and its cells.

The "natural history" of the disease needs to be completed by specific consideration of factors related to the microorganism, the host and the food matrix, insofar as they may affect development of health effects, their frequency and severity.

5.2 INFORMATION RELATED TO THE PATHOGEN

Basically, this information is analysed with a view to determining the characteristics of the pathogen that affect its ability to cause disease in the host. The analysis will take the biological nature of the pathogen into account (bacterial, viral, parasitic, prion) as well as the relevant mechanisms that cause illness (infectious, toxico-infectious, toxigenic, invasive or not, immune-mediated illness, etc.). In principle, the descriptive hazard characterization is applicable to all types of pathogens and all associated illnesses. In practice, by nature of the data collected, the focus will be on acute effects, associated with single exposures rather than long-term effects associated with chronic exposure. Note that the possible interaction between repeated single exposures (e.g. the development of acquired immunity) is an integral part of the descriptive characterization.

The ability of a pathogen to cause disease is influenced by many factors (Table 2). Some of these factors relate to the intrinsic properties of the pathogen, such as phenotypic and genetic characteristics that influence virulence and pathogenicity, and host specificity. The characteristics of the pathogen that determine its ability to survive and multiply in food and water, based on its resistance to processing conditions, are critical components of MRA, with reference to both exposure assessment and hazard characterization. Ecology, strain variation, infection mechanisms and potential for secondary transmission may also be considered, depending on the biology of the microorganism and on the context of the hazard characterization, such as the scenario that has been delineated during the problem formulation stage of a full risk assessment.

Table 2. Elements that might be included in characterization of the pathogen

Intrinsic properties of the pathogen (phenotypic and genetic characteristics)
Virulence and pathogenicity mechanisms
Pathological characteristics and disease caused
Host specificity
Infection mechanisms and portals of entry
Potential for secondary spread
Strain variability
Antimicrobial resistance and its effect on severity of disease

SOURCE: Adapted from ILSI, 2000.

When not already included in the characterization of the pathogen, specific consideration should be given to the intrinsic properties of the pathogen that influence infectivity, virulence and pathogenicity; their variability; and the factors that may alter the infectivity, virulence or pathogenicity of the microorganism under consideration. As a minimum, elements to be included in hazard characterization with regard to the pathogen are summarized in Table 2.

5.3 INFORMATION RELATED TO THE HOST

Host-related factors are the characteristics of the potentially exposed human population that may influence susceptibility to the particular pathogen, taking into account host intrinsic and acquired traits that modify the likelihood of infection or, most importantly, the probability of illness and its severity. Host barriers are multiple in number and pre-existing (innate); they are not all equally effective against pathogens. Each barrier component may have a range of effects depending on the pathogen, and many factors may influence susceptibility and severity. These are identified in Table 3.

Not all of the factors listed in Table 3 would be relevant, or important, for all pathogens. In all cases, however, an important issue in hazard characterization is to provide information on whom is at risk and on the stratification of the exposed population for relevant factors that influence susceptibility and severity.

5.4 INFORMATION RELATED TO THE MATRIX

The factors related to the food matrix are principally those that may influence the survival of the pathogen through the hostile environment of the stomach. Such effects may be induced by protection of the pathogen against physiological challenges, such as gastric acid or bile salts.

Table 3. Factors related to the host that may influence susceptibility and severity
Age
General health status, stress
Immune status
Underlying conditions, concurrent or recent infections
Genetic background
Use of medications
Pertinent surgical procedures
Pregnancy
Breakdown of physiological barriers
Nutritional status, bodyweight
Demographic, social, and behavioural traits
SOURCE: Adapted from ILSI, 2000.

These are related to the composition and structure of the matrix (e.g. highly buffered foods; entrapment of bacteria in lipid droplets). Alternatively, the conditions in the matrix may phenotypically affect the ability of the pathogen to survive the host barriers, such as increased acid tolerance of bacteria following pre-exposure to moderately acid conditions, or induction of stress-response by starvation in the environment. Stress conditions encountered during the processing or distribution of food and water may alter a pathogen's inherent virulence and its ability to resist the body's defence mechanisms. These potential matrix effects can be important elements in hazard characterization. The conditions of ingestion may also influence survival by altering the contact time between pathogens and barriers, e.g. initial rapid transit of liquids in an empty stomach. These factors are summarized in Table 4.

Table 4. Elements that may be included in characterization of the effect of the matrix on the pathogen-host relationship
Protection of the pathogen against physiological barriers
Induction of stress response
Effects on transport of pathogen through the gastrointestinal tract

5.5 DOSE-RESPONSE RELATIONSHIP

The final – and essential – element in the descriptive hazard characterization is the relationship, if any, between the ingested dose, infection and the manifestation and magnitude of health effects in exposed individuals.

Description of the dose-response relationship involves consideration of the elements or factors related to the pathogen, the host and the matrix, insofar as they may modulate the response to exposure. Where appropriate information is available, it also involves a discussion about the biological mechanisms involved, in particular whether a threshold, or a collaborative action of the pathogens, may be a plausible mechanism for any harmful effect, or whether a single pathogen may cause adverse effects under certain circumstances. Elements to be considered are listed in Table 5.

Where clinical or epidemiological data are available, discussion of the dose-response relationship will generally be based on such data. However, the quality and quantity of data available will affect

Table 5. Elements to be considered in describing the dose-response relationship
Organism type and strain
Route of exposure
Level of exposure (the dose)
Adverse effect considered (the response)
Characteristics of the exposed population
Duration – multiplicity of exposure
SOURCE: Adapted from ILSI, 2000.

the characterization. The strengths and limitations of the different types of data were addressed in Chapter 4. A specific difficulty is of obtaining data to characterize infection, or to characterize the translation of infection into illness and illness into different outcomes. In many cases, the analysis may only be able to describe a relationship between a dose and clinical illness. Other difficulties arise from several sources of variability, including variation in virulence and pathogenicity of the microorganisms, variation in attack rates, variation in host susceptibility, and type of vehicle, which modulates the ability of pathogens to affect the host. Therefore, it is essential that the dose-response analysis clearly identify what information has been utilized and how the information was obtained. In addition, the variability should be clearly acknowledged and the uncertainties and their sources, such as insufficient experimental data, should be thoroughly described.

6. DOSE-RESPONSE MODELLING

Concurrently with the descriptive analysis of clinical or epidemiological information or data, mathematical modelling has been advocated to provide assistance in developing a dose-response relationship, in particular when extrapolation to low doses is necessary. Mathematical models have been used for several decades in the field of toxicology. In the field of water and food microbiology, it is currently recognized that mathematical models may facilitate the dose-response assessment exercise, and provide useful information while accounting for variability and uncertainty. The assumptions on which current models are based, their use and possible limitations are carefully considered in the following sections.

The focus of these sections is on infectious and toxico-infectious pathogens, as this has been the area of most development. Some attention is given to other pathogens at the end of the chapter.

6.1 THE INFECTIOUS DISEASE PROCESS

The biological basis for dose-response models derives from major steps in the disease process as they result from the interactions between the pathogen, the host and the matrix. Figure 4 illustrates the major steps in the overall process, with each step being composed of many biological events. Infection and illness can be seen as resulting from the pathogen successfully passing multiple barriers in the host. These barriers are not all equally effective in eliminating or inactivating pathogens and may have a range of effects, depending on the pathogen and the individual. Each individual pathogen has some particular probability to overcome a barrier, which is conditional on the previous step(s) being completed successfully. The disease process as a whole and each of the component steps may vary by pathogen and by host. Pathogens and hosts can be grouped with regard to one or more components, but this should be done cautiously and transparently.

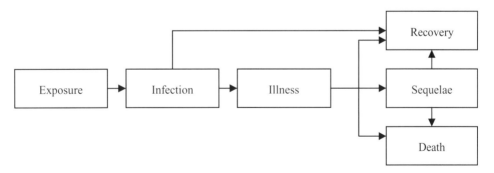

Figure 4. The major steps in the foodborne infectious disease process.

A dose-response model describes the probability of a specified response from exposure to a specified pathogen in a specified population, as a function of the dose. This function is based on empirical data, and will usually be given in the form of a mathematical relationship. The use of mathematical models is needed because:

- contamination of food and water usually occurs with low numbers or under exceptional circumstances; the occurrence of effects can not usually be measured by observational methods in the dose range needed, and hence models are needed to extrapolate from high doses or frequent events to actual exposure situations;

- pathogens in food and water are usually not randomly dispersed but appear in distinct clumps or clusters, which must be taken into account when estimating health risks; and

- experimental group sizes are limited, and models are needed, even in well controlled experiments, to distinguish random variation from true biological effects.

Plots of empirical datasets relating the response of a group of exposed individuals to the dose (often expressed as a logarithm) frequently show a sigmoid shape, and can be fitted by a large number of mathematical functions. However, when extrapolating outside the region of observed data, these models may predict widely differing results (cf. Coleman and Marks, 1998; Holcomb et al., 1999). It is therefore necessary to select between the many possible dose-response functions. In setting out to generate a dose-response model, the biological aspects of the pathogen-host-matrix interaction should be considered carefully. The model functions derived from this conceptual information should then be treated as *a priori* information. For more details, see Section 6.2.

6.1.1 Exposure

In general, biologically plausible dose-response models for microbial pathogens should consider the discrete (particulate) nature of organisms and should be based on the concept of infection from one or more "survivors" from an initial dose. Before proceeding, however, it is necessary to carefully consider the concept of "dose".

The concentration of pathogens in the inoculum is usually analysed by some microbiological, biochemical, chemical or physical method. Ideally, such methods would have 100% sensitivity and specificity for the target organism, but this is rarely the case. Therefore it may be necessary to correct the measured concentration for the sensitivity and specificity of the measurement method to provide a realistic estimate of the number of viable, infectious agents. The result may be greater or smaller than the measured concentration. Note that, in general, the measurement methods used to characterize the inoculum in a data set used for dose-response modelling will differ from the methods used to characterize exposure in a risk assessment model. These differences need to be accounted for in the risk assessment.

Multiplying the concentration of pathogens in the inoculum by the volume ingested, the mean number of pathogens ingested by a large group of individuals can be calculated. The actual number ingested by any exposed individual is not equal to this mean, but is a variable number that can be characterized by a probability distribution. It is commonly assumed that the pathogens are randomly distributed in the inoculum, but this is rarely the case. Compound distribution (or over-dispersion) may result from two different mechanisms:

- A 'unit" as detected by the measurement process (e.g. a colony-forming unit (CFU), a tissue culture infectious dose, or a Polymerase Chain Reaction (PCR) detectable unit) may, due to aggregation, consist of more than one viable, infectious particle. This is commonly observed for viruses, but may also be the case for other pathogens. The degree of clumping strongly depends on the methods used for preparing the inoculum.

- In a well-homogenized liquid suspension, unit doses will be more or less randomly distributed. If the inoculum consists of a solid or semisolid food matrix, however, spatial clustering may occur and result in over-dispersion of the inoculum. This aspect may differ between the data underlying the dose-response model and the actual exposure scenario.

The Poisson distribution is generally used to characterize the variability of the individual doses when pathogens are randomly distributed. Microorganisms have a tendency to aggregate in aqueous suspensions. In such cases, the number of "units" counted is not equal to the number of infectious particles but to the number of aggregates containing one or more infectious particles. In such cases, it is important to know whether the aggregates remain intact during inoculum preparation or in the gastrointestinal tract. Also, different levels of aggregation in experimental samples and in actual water or food products need to be accounted for.

6.1.2 Infection

Each individual organism in the ingested dose is assumed to have a distinct probability of surviving all barriers to reach a target site for colonization. The relation between the actual number of surviving organisms (the effective dose) and the probability of colonization of the host is a key concept in the derivation of dose-response models, as will be discussed later.

Infection is most commonly defined as a situation in which the pathogen, after ingestion and surviving all barriers, actively grows at its target site (Last, 1995). Infection may be measured by different methods, such as faecal excretion or immunological response. Apparent infection rates may differ from actual infection rates, depending on the sensitivity and specificity of the diagnostic assays. Infection is usually measured as a quantal response (presence or absence of infection by some criterion). The use of continuous-response variables (e.g. an antibody titre) may be useful for further development of dose-response models. Infections may be asymptomatic, where the host does not develop any adverse reactions to the infection, and clears the pathogens within a limited period of time, but infection may also lead to symptomatic illness.

6.1.3 Illness

Microbial pathogens have a wide range of virulence factors, and may elicit a wide spectrum of adverse responses, which may be acute, chronic or intermittent. In general, disease symptoms may result from either the action of toxins or damage to the host tissue. Toxins may have been preformed in the food or water matrix ("intoxication") or may be produced *in vivo* by microorganisms in the gut ("toxico-infection"), and may operate by different pathogenic mechanisms (e.g. Granum, Tomas and Alouf, 1995). Tissue damage may also result from a wide range of mechanisms, including destruction of host cells, invasion and inflammatory responses. For many foodborne pathogens, the precise pathogenic sequence of events is unknown, and is likely to be complex. Note that health risks of toxins in water (e.g. cyanobacterial toxins) usually relate to repeated exposures, and these require another approach, which resembles hazard characterization of chemicals.

Illness can basically be considered as a process of cumulative damage to the host, leading to adverse reactions. There are usually many different and simultaneous signs and symptoms of illness in any individual, and the severity of symptoms varies among pathogens and among hosts infected with the same pathogen. Illness is therefore a process that is best measured on a multidimensional, quantitative, continuous scale (number of stools passed per day, body temperature, laboratory measurements, etc.). In contrast, in risk assessment studies, illness is usually interpreted as a quantal response (presence or absence of illness), implying that the results depend strongly on the case definition. A wide variety of case definitions for gastrointestinal illness are used in the literature, based on a variable list of symptoms, with or without a specified time window, and sometimes including laboratory confirmation of etiological agents. This lack of standardization severely hampers integration of data from different sources.

6.1.4 Sequelae and mortality

In a small fraction of ill persons, chronic infection or sequelae may occur. Some pathogens, such as *Salmonella enterica* serotype Typhi, are invasive and may cause bacteraemia and generalized infections. Other pathogens produce toxins that may result not only in enteric disease but also in severe damage in susceptible organs. An example is haemolytic uraemic syndrome, caused by damage to the kidneys from Shiga-like toxins of some *Escherichia coli* strains. Complications may also arise by immune-mediated reactions: the immune response to the pathogen is then also directed against the host tissues. Reactive arthritis (including Reiter's syndrome) and Guillain-Barré syndrome are well known examples of such diseases. The complications from gastroenteritis normally require medical care, and frequently result in hospitalization. There may be a substantial risk of mortality in relation to sequelae, and not all patients may recover fully, but may suffer from residual symptoms, which may last a lifetime. Therefore, despite the low probability of complications, the public health burden may be significant. Also, there is a direct risk of mortality related to acute disease, in particular in the elderly, neonates and severely immunocompromised.

6.2 MODELLING CONCEPTS

Several key concepts are required for the formulation of biologically plausible dose-response models. These relate to:

- threshold vs non-threshold mechanisms;
- independent vs synergistic action; and

- the particulate nature of the inoculum.

Each of these concepts will be discussed below in relation to the different stages of the infection and disease process. Ideally, the dose-response models should represent the following series of conditional events: the probability of infection given exposure; the probability of acute illness given infection; and the probability of sequelae or mortality given acute illness.

In reality, however, the necessary data and concepts are not yet available for this approach. Therefore models are also discussed that directly quantify the probability of illness or mortality given exposure.

6.2.1 Threshold vs non-threshold mechanisms

The traditional interpretation of dose-response information was to assume the existence of a threshold level of pathogens that must be ingested in order for the microorganism to produce infection or disease. A threshold exists if there is no effect below some exposure level, but above that level the effect is certain to occur. Attempts to define the numerical value of such thresholds in test populations have typically been unsuccessful, although the concept is widely referred to in the literature as the "minimal infectious dose".

An alternative hypothesis is that, due to the potential for microorganisms to multiply within the host, infection may result from the survival of a single, viable, infectious pathogenic organism ("single-hit concept"). This implies that, no matter how low the dose, there is always, at least in a mathematical sense, and possibly very small, a non-zero probability of infection and illness. Obviously, this probability increases with the dose.

Note that the existence or absence of a threshold, at both the individual and population levels, cannot be demonstrated experimentally. Experimental data are always subject to an observational threshold (the experimental detection limit): an infinitely small response cannot be observed. Therefore, the question of whether a minimal infectious dose truly exists or merely results from the limitations of the data tends to be academic. A practical solution is to fit dose-response models that have no threshold (no mathematical discontinuity), but are flexible enough to allow for strong curvature at low doses so as to mimic a threshold-like dose-response.

The probability of illness given infection depends on the degree of host damage that results in the development of clinical symptoms. For such mechanisms, it seems to be reasonable to assume that the pathogens that have developed *in vivo* must exceed a certain minimum number. A non-linear relation may be enforced because the interaction between pathogens may depend on their numbers *in vivo*, and high numbers are required to switch on virulence genes (e.g. density dependent quorum-sensing effects). This concept, however, is distinct from a threshold for administered dose, because of the possibility, however small, that a single ingested organism may survive the multiple barriers in the gut to become established and reproduce.

6.2.2 Independent action vs synergistic action

The hypothesis of independent action postulates that the mean probability p per inoculated pathogen to cause (or help cause) an infection (symptomatic or fatal) is independent of the number of pathogens inoculated, and for a partially resistant host it is less than unity. In contrast, the hypotheses of maximum and of partial synergism postulate that inoculated pathogens cooperate so that the value of p increases as the size of the dose increases (Meynell and Stocker, 1957). Several experimental studies have attempted to test these hypotheses and the results have generally been consistent with the hypothesis of independent action (for a review, see Rubin, 1987).

Quorum sensing is a new area of research that is clearly of importance in relation to the virulence of some bacteria. It means that some phenotypic characteristics such as specific virulence genes are not expressed constitutively, but are rather cell-density dependent, using a variety of small molecules for cell-to-cell signalling, and are only expressed once a bacterial population has reached a certain density (De Kievit and Iglewski, 2000). While the biology of quorum sensing and response is still being explored, the nature of the effect is clear, it may be that some virulence factors are only expressed once the bacterial population reaches a certain size. The role of quorum sensing in the early stages of the infectious process has not been investigated in detail, and no conclusion can be drawn about the significance of quorum sensing in relation to the hypothesis of independent action. In particular, the role of interspecies and intraspecies communication is an important aspect. Sperandio et al. (1999) have demonstrated that intestinal colonization by enteropathogenic *E. coli* could be induced by quorum sensing of signals produced by non-pathogenic *E. coli* of the normal intestinal flora.

6.3 SELECTION OF MODELS

Specific properties in the data become meaningful only within the context of a model. Different models may, however, lead to different interpretations of the same data, and so a rational basis for model selection is needed. Different criteria may be applied when selecting mathematical models. For any model to be acceptable, it should satisfy the statistical criteria for goodness of fit. However, many different models will usually fit a given data set (for example, see Holcomb et al., 1999) and therefore goodness of fit is not a sufficient criterion for model selection. Additional criteria that might be used are conservativeness and flexibility.

Conservativeness can be approached in many different ways: "Is the model structure conservative?" "Are parameter estimates conservative?" "Are specific properties of the model conservative?" and so forth. It is not recommended to build conservativeness into the model structure itself. From a risk assessment perspective, a model should be restricted to describing the data and trying to discriminate the biological signal from the noise. Adding parameters usually improves the goodness of fit of a model, but using a flexible model with many parameters may result in greater uncertainty of estimates, especially for extrapolated doses. Flexible models and sparse datasets may lead to overestimation of the uncertainty, while a model based on strong assumptions may be too restrictive and lead to underestimation of the uncertainty in risk estimates.

It is recommended that dose-response models be developed based on a set of biologically plausible, mechanistic assumptions, and then to perform statistical analysis with those models that are considered plausible. Note that it is generally not possible to "work back", i.e. to deduce the assumptions underlying a given model formula. There is a problem of identifiability: the same functional form may result from different assumptions, while two (or more) different functional forms (based on different assumptions) may describe the same dose-response data equally well. This may result either in very different fitted curves if the data contains little information, or virtually the same curves if the data contain strong information. However, even in the last case, the model extrapolation may be very different. This means that a choice between different models or assumptions cannot be made on the basis of data alone.

6.3.1 Dose-infection models

The foregoing considerations lead us to the working hypothesis that, for microbial pathogens, dose-infection models based on the concepts of single-hit and independent action are regarded as scientifically most plausible and defensible. When the discrete nature of pathogens is also taken into account, these concepts lead to the single-hit family of models, as detailed in Box 1.

The single-hit models are a specific set of models in a broader class of mechanistic models. Haas, Rose and Gerba (1999) describe models that assume the existence of thresholds – whether constant or variable – for infection, i.e. some minimum number of surviving organisms larger than 1 is required for the infection to occur. Empirical (or tolerance distribution) models, such as the log-logistic, log-probit and Weibull(-Gamma) models, have also been proposed for dose-response modelling. The use of these alternative models is often motivated by the intuitive argument that single-hit models overestimate risks at low doses.

6.3.2 Infection-illness models

Currently, infection-illness models have received little attention and data available are extremely limited. Experimental observations show that the probability of acute illness among infected subjects may increase with ingested dose, but a decrease has also been found (Teunis, Nagelkerke and Haas, 1999), and often the data do not allow conclusions about dose dependence, because of the small numbers involved. Given this situation, constant probability (i.e. independent of the ingested dose) models, possibly stratified for subgroups in the population with different susceptibilities, seem to be a reasonable default. Together with ingested dose, illness models should take into account the information available on incubation times, duration of illness and timing of immune response, and should preferably measure illness as a multidimensional concept on continuous scales. There is no basis yet to model the probability of illness as a function of the numbers of pathogens that have developed in the host.

Box 1. Hit-theory models

Consider a host that ingests exactly one cell of a pathogenic microorganism. According to the single-hit hypothesis, the probability that this pathogen will survive all barriers and colonize the host has a non-zero value of p_m Thus, the probability of the host not being infected is $1-p_m$. If a second cell of the pathogen is ingested, and the hypothesis of independent action is valid, then the probability of the host not being infected is $(1-p_m)^2$. For n pathogens, the probability of not being infected is $(1-p_m)^n$. Hence, the probability of infection of a host that ingests exactly n pathogens can be expressed as:

$$P_{inf}(n; p_m) = 1 - (1-p_m)^n$$

Starting from this basic function, a broad family of dose-response models (hit-theory models) can be derived. The most frequently used models are the exponential and the Beta-Poisson models, which are based on further assumptions on the distribution of pathogens in the inoculum, and on the value of p_m. When the distribution of the organisms in the inoculum is assumed to be random, and characterized by a Poisson distribution, it can be shown (e.g. Teunis and Havelaar, 2000) that the probability of infection as a function of the dose is given by:

$$P_{inf}(D; p_m) = 1 - e^{-D.p_m}$$

where D is the mean ingested dose. If p_m is assumed to have a constant value r for any given host and any given pathogen, the simple exponential model results:

$$P_{inf}(D; r) = 1 - e^{-rD}$$

When $r.D \ll 1$, this formula is approximated by:

$$P_{inf}(D; r) \approx r.D$$

If the probability of starting an infection differs for any organism in any host, and is assumed to follow a beta-distribution, then:

$$P_{inf}(D; \alpha, \beta) = 1 - {}_1F_1(\alpha, \alpha + \beta, -D)$$

For $\alpha \ll \beta$ and $\beta \gg 1$, the Kummer confluent hypergeometric function ${}_1F_1$ is approximately equal to the Beta-Poisson formula:

$$P_{inf}(D; \alpha, \beta)) \approx 1 - (1 + \frac{D}{\beta})^{-\alpha}$$

When $\frac{\alpha}{\beta}.D \ll 1$, this formula is approximated by $P_{inf}(D; \alpha, \beta)) \approx \frac{\alpha}{\beta}.D$.

For both $\alpha \to \infty$ and $\beta \to \infty$, while $\frac{\alpha}{\beta} \to r$, the Beta-Poisson formula converts into the exponential model.

Other assumptions for n or p lead to other models. For example, spatial clustering of cells in the inoculum can be represented by a negative binomial distribution or any other contagious distribution. However, this has little effect on the shape of the dose-response relationship (Haas, Rose and Gerba, 1999) although the limiting curve for the confidence interval is affected (Teunis and Havelaar, 2000). It is also possible to model p_m as a function of covariables, such as immune status or age.

6.3.3 Dose-illness models

The default assumption of constant probability models for illness given infection lead to the conclusion that the only difference between dose-infection and dose-illness models is that the dose-illness models do not need to reach an asymptote of 1, but of P(ill|inf). They would essentially still belong to the family of hit-theory models.

6.3.4 Sequelae and mortality

Given illness, the probability of sequelae or mortality, or both, depends of course on the characteristics of the pathogen, but more importantly on the characteristics of the host. Sequelae or mortality are usually rare events that affect specific subpopulations. These may be identified by factors such as age or immune status, but increasingly genetic factors are being recognized as important determinants. As above, the current possibilities are mainly restricted to constant probability models. Stratification appears to be necessary in almost all cases where an acceptable description of risk grouping is available.

6.4 EXTRAPOLATION

6.4.1 Low dose extrapolation

Dose-response information is usually obtained in the range where the probability of observable effects is relatively high. In experimental studies using human or animal subjects, this is related to financial, ethical and logistical restrictions on group size. In observational studies, such as outbreak studies, low dose effects can potentially be observed directly, but in these studies only major effects can be distinguished from background variation. Because risk assessment models often include scenarios with low dose exposures, it is usually necessary to extrapolate beyond the range of observed data. Mathematical models are indispensable tools for such extrapolations, and many different functional forms have been applied. Selection of models for extrapolation should primarily be driven by biological considerations, and only subsequently by the available data and their quality. The working hypotheses of no-threshold and independent action lead to a family of models that is characterized by linear low dose extrapolations on the log/log scale, or even on the arithmetic scale. That is, in the low dose range, the probability of infection or disease increases linearly with the dose. On the log-scale, these models have a slope of 1 at low doses. Some examples include:

- The exponential model $P = r.D$

- Beta-Poisson model $P = (\alpha/\beta).D$

- The hypergeometric model $P = \{\alpha/(\alpha+\beta)\}.D$

where D = mean ingested dose and r, α and β are model parameters. Note that if $\alpha > \beta$, the risk of infection predicted by the Beta-Poisson model is larger than the risk of ingestion, which is not biologically plausible. This highlights the need to carefully evaluate the appropriateness of using this simplified model for analysing dose-response data.

6.4.2 Extrapolation in the pathogen-host-matrix triangle

Experimental datasets are usually obtained under carefully controlled conditions, and the data apply to a specific combination of pathogen, host and matrix. In actual exposure situations, there is more variability in each of these factors, and dose-response models need to be generalized. Assessing such variability requires the use of multiple datasets that capture the diversity of human populations, pathogen strains and matrices. Failure to take such variation into account may lead to underestimation of the actual uncertainty of risks.

When developing dose-response models from multiple datasets, one should use all of the data that is pertinent. There is currently no way of determining which data source is best. This requires that the risk assessor make choices. Such choices should be based on objective scientific arguments to the maximum possible extent, but will inevitably include subjective arguments. Such arguments should be discussed with the risk manager and their significance and impact for risk management considered. The credibility of dose-response models increases significantly if dose-response relations derived from different data sources are consistent, especially if the data are of varying types.

When combining data from different sources, a common scale on both axes is needed. This often requires adjusting the reported data to make them comparable. For dose, test sensitivity, test specificity, sample size, etc., need to be taken into account. For response, a consistent case definition is needed or the reported response needs to be adjusted to a common denominator (e.g. infection × conditional probability of illness given infection). Combining data from different sources within a single (multilevel) dose-response model requires thorough statistical skills and detailed insight into the biological processes that generated the data. An example is the multilevel dose-response model that has been developed for different isolates of *Cryptosporidum parvum* (Teunis et al., 2002a). The issue of combining data from different outbreak studies is discussed in the FAO/WHO risk assessments of *Salmonella* in eggs and broiler chickens (FAO/WHO, 2002a).

Dose-response relations where an agent only affects a portion of the population may require that subpopulation to be separated from the general population in order to generate meaningful results. Using such stratified dose-response models in actual risk assessment studies requires that the percentage of the population that is actually susceptible can be estimated. Consideration of such subpopulations appears to be particularly important when attempting to develop dose-response relations for serious infections or mortality. However, it would also be pertinent when considering an agent for which only a portion of the population can become infected.

Stratified analysis can also be useful when dealing with seemingly outlying results, which may actually indicate a subpopulation with a different response. Removal of one or more outliers corresponds to removing (or separately analysing) the complete group from which the outlying results originated. Where a specific reason for the separation cannot be identified, there should be a bias toward being inclusive in relation to the data considered. Any elimination of the data should be clearly communicated to ensure the transparency of the assessment.

A particular and highly relevant aspect of microbial dose-response models is the development of specific immunity in the host. Most volunteer experiments have been conducted with test subjects selected for absence of any previous contact with the pathogen, usually demonstrated by absence of specific antibodies. The actual population exposed to foodborne and waterborne pathogens will usually be a mixture of totally naive persons and persons with varying degrees of protective immunity. No general statements can be made on the impact of these factors. This is strongly dependent on the pathogen and the host population. Some pathogens, such as many childhood diseases and the hepatitis A virus, will confer lifelong immunity upon first infection whether clinical or subclinical, whereas immunity to other pathogens may wane within a few months to a few years, or may be evaded by antigenic drift. At the same time, exposure to non-pathogenic strains may also protect against virulent variants. This principle is the basis for vaccination, but has also been demonstrated for natural exposure, e.g. to non-pathogenic strains of *Listeria monocytogenes* (Notermans et al., 1998). The degree to which the population is protected by immunity depends to a large extent on the general hygienic situation. In many developing countries, large parts of the population have built up high levels of immunity, and this is thought to be responsible for lower incidence or less serious forms of illness. Some examples are the predominantly watery form of diarrhoea by *Campylobacter* spp. infections in children and the lack of illness from this organism in young adults in developing countries. The apparent lack of *E. coli* O157:H7-related illness in Mexico has been explained as the result of cross-immunity following infections with other *E. coli*, such as enteropathogenic *E. coli* strains that are common there. In contrast, in the industrialized world, contact with enteropathogens is less frequent and a larger part of the population is susceptible. Obviously, age is an important factor in this respect.

Incorporating the effect of immunity in dose-response models has as of yet received little attention. The absence of accounting for immunity in dose-response models may complicate interpretations, and comparisons among places. This is particularly likely to be a problem with common infections such as *Campylobacter* spp., *Salmonella* spp. and *E. coli*. Immunity may affect the probability of infection, the probability of illness given infection, or the severity of illness. There are currently only few data available on which to base model development. Where such data are available, a simple and possibly effective option would be to resort to stratified analysis and divide the population into groups with different susceptibility (see, for example, FDA/USDA/CDC, 2001). Recently, experimental work on infection of volunteers having different levels of acquired immunity to *Cryptosporidium parvum* was analysed with a dose-response model that includes the effects of immunity (Teunis et al., 2002b).

6.5 FITTING DOSE RESPONSE MODELS TO DATA

First and foremost, like other parts of the risk assessment process, model-fitting procedures should be reported clearly and unambiguously, for transparency and to allow reproduction.

6.5.1 Fitting method

Likelihood-based methods are preferable. The approach taken depends on the kinds of data that are available, and the presumed stochastic variation present. For instance, for binary data, model fitting should be performed by writing down the appropriate binomial likelihood function. For a dose-response function $f(D;\theta)$ with parameter vector θ, the likelihood function for a set of observations is:

$$l(\theta) = \prod_i [f(D_i;\theta)]^{k_i} [1 - f(D_i;\theta)]^{n_i - k_i}$$

where the product is taken over all dose groups, with index i. At dose D_i, a number n_i of subjects is exposed, and k_i are infected. Fitting consists of finding parameter values that maximize this function, hence the term: maximum likelihood parameter values. Optimization may require special care, since many dose-response models are essentially non-linear. Most technical mathematics systems, such as Matlab, Mathematica or Gauss, or statistical systems, such as SAS, Splus, or R, provide procedures for non-linear optimization.

Haas (1983) and Haas, Rose and Gerba (1999) provide specific technical information on how to fit dose-response models. A general overview can be found in any textbook on mathematical statistics, such as Hogg and Craig (1994). McCullagh and Nelder (1989) is the definitive source for the statistical methods involved, and many dose-response models can be written as generalized linear models (but not the exact single-hit model – see Teunis and Havelaar, 2000). Vose (2000) is a valuable resource for a general description of mathematical and statistical methods in risk assessment.

6.5.2 Selection of the best fitting model or models

When the likelihood function of a model is available, model testing can be done by calculating likelihood ratios. Goodness of fit may be assessed against a likelihood supremum – a model with as many degrees of freedom as there are data (i.e. dose groups). For instance, for binary responses, a likelihood supremum may be calculated by inserting ratios of positive responses to numbers of exposed subjects into the binomial likelihood (McCullagh and Nelder, 1989):

$$l_{\text{sup}} = \prod_i \left(\frac{k_i}{n_i} \right)^{k_i} \left(1 - \frac{k_i}{n_i} \right)^{n_i - k_i}$$

The deviance, -2 × the difference in log-likelihood, can be approximated as a chi-square variate, with degrees of freedom equal to the number of dose groups minus the number of model parameters.

The same method can be used for model ranking. To compare two models, one starts by calculating maximum likelihoods for both models, and then determining their deviance (-2 × the difference in log-likelihoods). This deviance can now be tested against chi-square with degrees of freedom equal to the difference in numbers of parameters of the two models, at the desired level of significance (Hogg and Craig, 1994).

The chi-square approximation is asymptotically correct for large samples. In addition to this, the likelihood ratio test is only valid for models that are hierarchically nested, meaning that the more general model can be converted to the less general one by parameter manipulation. Model complexity may be addressed by using an information criterion, such as the Akaikes Information Criterion (AIC), instead of the likelihood ratio. This penalizes parameter abundance, to balance goodness of fit against parameter parsimony (i.e. the minimum number of parameters necessary).

More generally valid are Bayesian methods, allowing comparison among any models, not only nested ones. Goodness of fit can be compared with Bayes factors, and there is also a corresponding information criterion: the Bayesian Information Criterion (BIC) (Carlin and Louis, 1996).

6.5.3 Uncertainty analysis

Determination of parameter uncertainty is indispensable. Categories of methods that can be applied include:

- *Likelihood-based methods* The (log-)likelihood function as a chi-square deviate can be used to construct confidence intervals for parameters. For more than one parameter, the resulting uncertainty in the dose-response model cannot be calculated in a straightforward manner (Haas, Rose and Gerba, 1999).

- *Bootstrapping* Bootstrapping involves the generation of replicate data by means of re-sampling (Efron and Tibshirani, 1993). For instance, for binary data, replicates can be generated by random sampling from a binomial distribution at each dose, with number of trials equal to the number of exposed subjects, and probability the fraction of infected over exposed subjects (Haas, Rose and Gerba, 1999; Medema et al., 1996). The model can then be fitted to each of these replicate data sets, thereby producing a random sample of parameter estimates, one for each replicate. These may subsequently be employed to construct a confidence range for the dose response relation, or to assess the uncertainty at a given dose.

- *Markov chain Monte Carlo methods* (MCMC) Adaptive rejection sampling methods are a powerful and efficient means of sampling from posterior distributions, especially when models with many parameters need to be analysed. Working within a Bayesian framework avoids many of the implicit assumptions that restrict the validity of classical likelihood methods, and so MCMC methods are rapidly becoming more common. For instance, most data sets used for dose-response analysis are very small, containing only a few dose groups with a few exposed subjects. Current interest in these methods has also increased the availability of ready-to-use tools (Gilks, Richardson and Spiegelhalter, 1996).

Most dose-response analyses for pathogenic microorganisms to date have only considered binary responses (infected or not; ill or not). Since, in such a context, each dose group may contain a mixture of responses, analysis of the heterogeneity in the response (segregation of uncertainty and variation) is not possible. Modelling infection as the amount of pathogens excreted, or elevation of one or more immune variables, or combinations of these, provides better opportunities for addressing heterogeneity within the host population and the pathogen population, and their segregation.

7. REVIEW

7.1 VALIDATION OF DOSE-RESPONSE MODELS

Model validation can be defined as demonstrating the accuracy of the model for a specified use. Within this context, accuracy is the absence of systematic and random error – in metrology they are commonly known as trueness and precision, respectively. All models are, by their nature, incomplete representations of the system they are intended to model, but, in spite of this limitation, models can be useful. General information on working with mathematical models can be found in various theoretical and applied textbooks. Doucet and Sloep (1992) give a very good introduction to model testing. These authors discriminate between model confirmation (i.e. shown to be worthy of our belief; plausible) and model verification (i.e. shown to be true). McCullagh and Nelder's book on linear models (1989) is a valuable resource on statistical modelling methods, and describes some general principles of applying mathematical models, underlining three principles for the modeller:

1. All models are wrong, but some are more useful than others.

2. Do not fall in love with one model to the exclusion of others.

3. Thoroughly check the fit of a model to the data.

In addition Law and Kelton (2000) in addressing the issue of building valid, credible and appropriately detailed simulation models consider techniques for increasing model validity and credibility. It is worth noting however that some models cannot be fully validated, but components or modules of the model can be validated on an individual basis. Dee (1995) has identified four major aspects associated with model validation, as follows:

1. Conceptual validation

2. Validation of algorithms

3. Validation of software code

4. Functional validation

These are described below. The issue of validation is further addressed in the FAO/WHO guidelines for Exposure assessment of microbiological hazards in foods and Risk characterization of microbiological hazards in foods.

Conceptual validation concerns the question of whether the model accurately represents the system under study. Was the simplification of the underlying biological process in model steps realistic, i.e. were the model assumptions credible? Usually, conceptual validation is largely qualitative and is best tested against the opinion of experts with different scientific backgrounds. Different models with various conceptual bases can be tested against each other within a Bayesian framework, using Bayes factors, or some information criterion. Experimental or observational data in support of the principles and assumptions should be presented and discussed. The modelling concepts described in Chapter 6 are a minimum set of assumptions representing the consensus opinion of a broad group of experts who contributed to these guidelines. These are based on mechanistic reasoning, and are supported by some experimental evidence. As such, they are considered to be currently the best basis for dose-response modelling studies.

Algorithm validation concerns the translation of model concepts into mathematical formulae. It addresses questions such as: Do the model equations represent the conceptual model? Under which conditions can simplifying assumptions be justified? What effect does the choice of numerical methods for model solving have on the results? and: Is there agreement among the results from use of different methods to solve the model? For microbiological dose-response models, these questions relate both to the adequacy of models (discussed in Section 6.3) for describing the infection and illness processes, and to the various choices outlined in here. Is it valid to assume a constant probability of infection for each pathogen in each host? Can the approximate Beta-Poisson model be used or is it necessary to use the exact hypergeometric model? A powerful method to evaluate the effects of numerical procedures is to compare the results of different methods used to estimate parameter uncertainty, such as overlaying parameter samples obtained by Monte Carlo or bootstrap procedures with likelihood-based confidence intervals. Graphical representation of the results can be useful, but must be used with care.

Software code validation concerns the implementation of mathematical formulae in computer language. Good programming practice (i.e. modular and fully documented) is an essential prerequisite. Specific points for attention are the possible effects of machine precision and software-specific factors on the model output. Internal error reports of the software are important sources of information, as well as evaluation of intermediate output. For dose-response models, it is advisable to check the results of a new software implementation against previously published results.

Functional validation concerns checking the model against independently obtained observations. Ideally, it is evaluated by obtaining pertinent real-world data, and performing a statistical comparison of simulated outcomes and observations. This requires more detailed information than is usually available. It may be possible to compare results from risk assessment studies with independently obtained epidemiological estimates of disease incidence. Such data cannot validate the dose-response model *per se,* but may produce valuable insights. Most studies to date have considered that a range check of estimated risks and observed incidences was sufficient "validation" of the model. However, the very nature of risk estimates (estimated probabilities) allows their use as a likelihood function to do a more formal test of adequacy.

Credibility of results can also be established by demonstrating that dose-response relationships computed from different data sources are consistent. For example, a dose-response relationship developed from a human feeding study may be validated against outbreak data or data from national surveillance of foodborne diseases. When making such comparisons, the different nature of hosts, pathogens and matrices must be accounted for. Thus, different sources of data may either be useful for model validation, or to provide a better basis for generalization.

7.2 PEER AND PUBLIC REVIEW

The process used to develop the results can improve the credibility of hazard characterization results. Peer and public review of results is a fundamental part of the process. Interdisciplinary interaction is essential to the process of risk assessment, and should be extended to the review process. Experts in the biological processes involved should review the basic concepts and underlying assumptions used in a hazard characterization. Furthermore, statistical experts should review the data analysis and presentation of model fitting results. A critical factor in obtaining a good peer review is to provide an accessible explanation of the mathematics, such that non-mathematicians can work their way through the concepts and details of the assessment.

Critical evaluation of a hazard characterization process is a demanding task that requires highly specialized experts. Therefore, adequate resources for the peer review process should be made available as an integral part of the project plan. The results of the peer review process should be accessible to all interested parties, including a statement on how comments were incorporated in the final version of the document and, if relevant, reasons why specific comments were not accepted.

The public review process serves two main purposes. First, it allows all stakeholders in a hazard characterization to critically review the assumptions made, and their effect on the risk assessment results. Second, it allows for evaluation of the completeness of the descriptive information and datasets used for the hazard characterization.

8. PRESENTATION OF RESULTS

Transparency requires that results are both accessible and understandable. Consequently, consideration should be given to the form of the presentation of hazard characterization results for both technical and non-technical audiences. A recommended schema for presentation of hazard characterization results is given in Appendix A

A quantitative interpretation of available information is preferable, and methods available for quantitative evaluation of subjective information should be assessed (Cooke, 1991). Even if there is no means for evaluating qualitative or subjective data, the results of the quantitative development of a dose-response relationship must be discussed and reconciled with additional epidemiological data in order to put it in context with the larger body of information. This is important in order to obtain acceptance of the results in the wider public health community.

When a quantitative approach is used, these guidelines recommend a biased neutral approach to hazard characterization, i.e. the best estimates of dose-response should be presented, together with attendant uncertainty. This approach requires distinction between variability and uncertainty. Variability is the observed differences attributable to true heterogeneity of diversity in a population or exposure parameter. In hazard characterization and MRA, variability cannot be reduced, only more precisely characterized. Uncertainty is ambiguity arising from lack of knowledge about specific factors, parameters or models. In the case of quantitative hazard characterizations, the most likely parameter values and their uncertainty should also be clearly communicated.

A sample of parameter values is usually necessary for use in risk assessment studies, and should, if possible, be made available in digital form, possibly through the Internet. Tables summarizing the results should include confidence intervals as well as characteristic or most-likely values. Graphical presentation of results can be useful in presenting uncertainty. A commonly used presentation graphs the response as a function of the dose (often in logarithmic form), with data points representing observed data and lines showing the fitted model (e.g. best fitting curve and the 5th and 95th percentile limits). However, for the small populations used in most experiments, the fraction of responses can only assume a limited number of discrete values (e.g. 0, 50 or 100% if two subjects were studied), which may unjustly suggest a poor fit of the model. Data clouds of parameter values or line graphs with error bars may be also helpful. The information that should be presented includes any assumptions made, summary statistics, and references for the data and methods.

A formalized means of presenting and disseminating results of hazard characterization (and MRAs) (e.g. a clearinghouse for dose-response functions and documentation) could speed the application of risk assessment.

Risk assessors should consider the effects of uncertainty from a number of sources – including model uncertainty, measurement uncertainty and extrapolation uncertainty – on the results of their model. Also, the sensitivity of the analysis to various assumptions and decisions made should be carefully evaluated and fully documented. The clear presentation of results could improve the guidance provided for the design of effective public health interventions for the pathogen of interest.

Hazard characterization and MRA results should be shared to the maximum extent possible in order to facilitate work in MRA. Dose-response relationships developed in one context will not be directly adaptable to another and will require careful consideration of differences in populations. Differences in age and immune status of the population, pathogen virulence and other variables will alter the dose-response relationship and these variables must be carefully considered. Adaptation of dose-response relationships outside of the population for which they were developed will require consideration of differences affecting the relationship and will require validation with data from the new population of interest.

9. REFERENCES CITED

CAC [Codex Alimentarius Commission]. 1999. Principles and guidelines for the conduct of Microbiological Risk Assessment. Document no. CAC/GL-30.

Carlin, B.P., & Louis, T.A. 1996. *Bayes and empirical Bayes methods for data analysis.* Monographs on Statistics and Applied Probability, No. 69. London: Chapman and Hall.

Coleman, M., & Marks, H. 1998. Topics in dose-response modelling. *Journal of Food Protection*, **61**(11): 1550–1559.

Cooke, R.M. 1991. *Experts in uncertainty.* New York, NY: Oxford University Press.

Dee, D.P. 1995. A pragmatic approach to model validation. Pp. 1–13, *in:* D.R. Lynch and A.M. Davies (eds). *Quantitative skill assessment of coastal ocean models.* Washington, DC: AGU.

De Kievit, T.R., & Iglewski, B.H. 2000. Bacterial quorum sensing in pathogenic relationships. *Infection and Immunity*, **68**: 4839–4849.

Doucet, P., & Sloep, P.B. 1992. *Mathematical modelling in the life sciences.* Chichester, England: Ellis Horwood Limited. [In particular, see Chapter 13 – Working with models, and Chapter 14 – Constructing models.]

Efron, B., & Tibshirani, R.J. 1993. *An introduction to the bootstrap.* Monographs on Statistics and Applied Probability, No. 57. London: Chapman and Hall.

FAO/WHO. 2002a. Risk assessments of *Salmonella* in eggs and broiler chickens. *Microbiological Risk Assessment Series,* Nos.1 and 2. WHO, Geneva, and FAO, Rome. Available on the Web at either www.fao.org/es/esn/food/risk_mra_riskassessment_salmonella_en.stm or www.who.int/fsf/mbriskassess/index.htm.

FAO/WHO. 2002b. Principles and Guidelines for incorporating microbiological risk assessment in the development of food safety standards, guidelines and related texts. Report of a joint FAO/WHO Expert Consultation, Kiel, Germany, 18–22 March 2002.

FDA/USDA/CDC [Food and Drugs Administration/United States Department of Agriculture/Center for Disease Control]. 2001. Draft assessment of the relative risk to public health from foodborne *Listeria monocytogenes* among selected categories of ready-to-eat foods. Available on the Web at www.foodsafety.gov/~dms/lmrisk.html.

Fewtrell, L., & Bartram, J. (eds). 2001. *Water quality: guidelines, standards and health. Assessment of risks and risk management for water-related infectious disease.* London: IWA Publishing.

Gilks, W.R., Richardson, S., & Spiegelhalter, D.J. 1996. *Markov chain Monte Carlo in practice.* London: Chapman and Hall.

Granum, P.E., Tomas, J.M., & Alouf, J.E. 1995. A survey of bacterial toxins involved in food poisoning: a suggestion for bacterial food poisoning toxin nomenclature. *International Journal of Food Microbiology*, **28**: 129–144.

Haas, C.N. 1983. Estimation of risk due to low doses of microorganisms: a comparison of alternative methodologies. *American Journal of Epidemiology*, **118**: 573–582.

Haas, C.N., Rose, J.B., & Gerba, C.P. 1999. *Quantitative microbial risk assessment.* New York, NY: John Wiley and Sons.

Havelaar, A.H., Heisterkamp, S.H., Hoekstra, J.A., & Mooijman, K.A. 1993. Performance characteristics of methods for the bacteriological examination of water. *Water Science Technology*, **27**: 1–13.

Havelaar, A.H., De Wit, M.A.S., Van Koningsveld, R., & Van Kempen, E. 2000. Health burden in the Netherlands due to infection with thermophilic *Campylobacter* spp. *Epidemiology and Infection,* **125**: 505–522.

Hogg, R.V., & Craig, A.T. 1994. *Introduction to mathematical statistics.* 5th ed. Upper Saddle River, NJ: Prentice Hall.

Holcomb, D.L., Smith, M.A., Ware, G.O., Hung, Y.C., Brackett, R.E., & Doyle, M.P. 1999. Comparison of six dose-response models for use with food-borne pathogens. *Risk Analysis*, **19**: 1091–1100.

ILSI [International Life Sciences Institute]. 2000. *Revised framework for microbial risk assessment.* Washington, D.C: ILSI Risk Science Institute Press.

Last, J.M. (ed). 1995. *A dictionary of epidemiology.* 3rd ed. New York, NY: Oxford University Press.

Law, A.M. and Kelton W.D. 2000. *Simulation Modeling and analysis.* 3rd edition, McGraw Hill, New York.

McCullagh, P., & Nelder, J.A. 1989. *Generalized linear models.* 2nd ed. London: Chapman & Hall. [In particular, see Chapter 1 – Introduction.]

Medema, G.J., Teunis, P.F.M., Havelaar, A.H., & Haas, C.N. 1996. Assessment of the dose-response relationship of *Campylobacter jejuni*. *International Journal of Food Microbiology*, **30**: 101–111.

Merrel, D.S., Butler, S.M., Qadri, F., Dolganov, N.A., Alam, A., Cohen, M.B., Calderwood, S.B., Schoolnik, G.K., & Camilla, A. 2002. Host-induced epidemic spread of the cholera bacterium. *Nature*, **417**: 642–645.

Meynell, G.G., & Stocker, B.A.D. 1957. Some hypotheses on the aetiology of fatal infections in partially resistant hosts and their application to mice challenged with *Salmonella paratyphi-B* or *Salmonella typhimurium* by intraperitoneal injection. *Journal of General Microbiology,* **16**: 38–58.

Notermans, S., Dufrenne, J., Teunis, P., & Chakraborty, T. 1998. Studies on the risk assessment of *Listeria monocytogenes*. *Journal of Food Protection*, **61**: 244–248.

Rubin, L.G. 1987. Bacterial colonization and infection resulting from multiplication of a single organism. *Review of Infectious Disease*, **9**: 488–493.

Sperandio, V., Mellies, J.L., Ngyuen, W., Shin, S., & Kaper, J.B. 1999. Quorum sensing controls expression of the type III secretion gene transcription and protein secretion in enterohemorrhagic and enteropathogenic *Escherichia coli*. *Proceedings of the National Academy of Science*, **96**: 15196–15201.

Teunis, P.F.M., & Havelaar, A.H. 2000. The Beta-Poisson model is not a single-hit model. *Risk Analysis,* **20**: 513–520.

Teunis, P.F.M., Chappell, C.L., & Ockhuysen, P.C. 2002a. *Cryptosporidium* dose response studies: variation between isolates. *Risk Analysis*, **22**: 175–183.

Teunis, P.F.M., Chappell, C.L., & Ockhuysen, P.C. 2002b. *Cryptosporidium* dose response studies: variation between hosts. *Risk Analysis*, **22**: 475–485.

Teunis, P.F.M., Nagelkerke, N.J.D., & Haas, C.N. 1999. Dose response models for infectious gastroenteritis. *Risk Analysis*, **19**: 1251–1260.

Vose, D. 2000. *Risk analysis. A quantitative guide.* 2nd ed. Chicester, UK: John Wiley and Sons.

WHO. 2000a. The World Health Report. Health systems: improving performance. Geneva: WHO.

WHO. 2000b. The interaction between assessors and managers of microbiological hazards in foods. Report of a WHO Expert Consultation. Kiel, Germany, 21–23 March 2000.

OUTLINE OF INFORMATION TO INCLUDE
IN A HAZARD CHARACTERIZATION

A suggested outline of the format and information to be included in a hazard characterization is provided below for reference purposes. Consideration of information to include starts with a summary of the host, pathogen and food matrix factors, and how these affect the likelihood of disease. The specific information included must be tailored to the purpose of the exercise and the pathogen commodity combination under consideration. Inclusion of a fitted dose-response curve depends on the quality of data available and the goodness of fit.

1. Description of the pathogen, host and food matrix factors and how these affect the disease outcome.

   1.1 Characteristics of the pathogen.

      1.1.1 Infectivity, virulence or pathogenicity, and disease mechanism.

      1.1.2 Genetic factors (e.g. antimicrobial resistance and virulence factors).

   1.2 Characteristics of the host or host population.

      1.2.1 Immunity status.

      1.2.2 Age, sex and ethnic group.

      1.2.3 Health behaviours.

      1.2.4 Physiological status.

      1.2.5 Genetic and environmental factors.

   1.3 Characteristics of the food matrix.

      1.3.1 Fat and salt content.

      1.3.2 pH and water activity.

      1.3.3 Processing related to stresses on microbial populations.

2. Public health outcomes.

   2.1 Manifestations of disease

   2.2 Rationale for the biological end-points modelled.

3. Dose-response relationship.

   3.1 Summary of available data.

Appendix B

GLOSSARY

Notes: (1) Sources are [numbered] where appropriate and listed at the end of the appendix.

(2) Some definitions are general but some are relevant to one specific discipline and in another discipline may be defined differently. In some cases, definitions are associated with a particular discipline, and these are here indicated as: (MRA) for microbiological risk assessment; (disease) for the infectious disease process; and (statistics) for statistical terminology.

Accuracy Degree of agreement between average predictions of a model or the average of measurements and the true value of the quantity being predicted or measured.

Adverse effect Change in morphology, physiology, growth, development or life span of an organism which results in impairment of functional capacity or impairment of capacity to compensate for additional stress or increase in susceptibility to the harmful effects of other environmental influences. Decisions on whether or not any effect is adverse require expert judgement. [2]

Akaikes Information Criterion (AIC) and **Bayesian Information Criterion** (BIC) These are criteria that are used in model selection to select the best model from a set of plausible models. One model is better than another model if it has a smaller AIC (or BIC) value. AIC is based on Kullback-Leibler distance in information theory, and BIC is based on integrated likelihood in Bayesian theory. If the complexity of the true model does not increase with the size of the data set, BIC is the preferred criterion, otherwise AIC is preferred. [18]

Aleatory uncertainty Aleatory is of or pertaining to natural or accidental causes and cannot be explained with mechanistic theory. Generally interpreted to be the same as *stochastic variability*.

Attack rate The proportion of an exposed population at risk that becomes infected or develops clinical illness during a defined period of time. [11]

Asymptomatic Showing or causing no symptoms (a symptom is any subjective evidence of disease or of a patient's condition, i.e. such evidence as perceived by the patient; a change in a patient's condition indicative of some bodily or mental state. Note that a symptom is different from a sign, which is any objective evidence of a disease, i.e. such evidence that is perceptible to the examining physician, as opposed to the subjective sensations (symptoms) of the patient. [12]

Bayesian inference Inference is using data to learn about some uncertain quantity. Bayes' theorem describes how to update a prior distribution about the uncertain quantity using a model (expressing likelihood of observed data) to obtain a posterior distribution. Bayesian inference allows incorporation of prior beliefs and can handle problems with insufficient data for frequentist inference.

Bayesian Information Criterion (BIC) See: *Akaikes Information Criterion* (AIC)

Bayesian methods These are an approach – founded on Bayes' Theorem – that forms one of the two flows of statistics. *Bayesian inference* is very strong when only subjective data is available and is useful for using data to improve one's estimate of a parameter.

Bias A term which refers to how far the average statistic lies from the parameter it is estimating, that is, the error which arises when estimating a quantity. It is also referred to as "systematic error". It is the difference between the mean of a model prediction or of a set of measurements and the true value of the quantity being predicted or measured. Errors from chance will cancel each other out in the long run; those from bias will not (statistics). [6]

Bootstrap A numerical method – also referred to as **Bootstrap simulation** – for inferring sampling distributions and confidence intervals for statistics of random variables. The methodology to estimate uncertainty involves generating subsets of the data on the basis of random sampling with replacements as the data are sampled. Such re-sampling means that each datum is equally represented in the randomization scheme (statistics). [7]

Case definition The case definition is a standard set of criteria for deciding whether an individual should be classified as having the health condition of interest. [15]

Confidence interval A range of values inferred or believed to enclose the actual or true value of an uncertain quantity with a specified degree of probability. Confidence intervals may be inferred based upon *sampling distributions* for a statistic.

Contagious distribution A probability distribution describing a stochastic process consisting of a combination of two or more processes. Also referred to as a "mixture distribution" (statistics).

Controllable variability Sources of heterogeneity of values of time, space or different members of a population that can be modified in part – in principle, at least – by intervention, such as a control strategy. For example, variability in the time and temperature history of food storage among storage devices influences variability in pathogen growth among food servings and in principle could be modified through a control strategy. For both population and individual risk, controllable variability is a component of overall variability.

Data quality objective Expectations or goals regarding the precision and accuracy of measurements, inferences from data regarding distributions for inputs, and predictions of the model.

Dose The amount of a pathogen that enters or interacts with an organism. [11]

Dose-response assessment The determination of the relationship between the magnitude of exposure (dose) to a chemical, biological or physical agent and the severity and/or frequency of associated adverse health effects (response) (MRA). [1]

Expert judgement   Judgement involves a reasoned formation of opinions. An expert is someone with special knowledge or experience in a particular problem domain. Expert judgement is documented and can be explained to satisfy outside scrutiny.

Exposure assessment The qualitative and/or quantitative evaluation of the likely intake of biological, chemical and physical agents via food, as well as exposure from other sources if relevant (MRA). [1]

Food Any substance, whether processed, semi-processed or raw, which is intended for human consumption, and includes drink, chewing gum and any substance that has been used in the manufacture, preparation or treatment of "food", but excludes cosmetics or tobacco, or substances used only as drugs. [1]

Goodness of fit   The statistical resemblance of real data to a model, expressed as a strength or degree of fit of the model (statistics). [8]

Goodness-of-fit test   A procedure for critiquing and evaluating the potential inadequacies of a probability distribution model with respect to its fitness to represent a particular set of observations.

Hazard A biological, chemical or physical agent in, or condition of, food with the potential to cause an adverse health effect (MRA). [1]

Hazard characterization The qualitative and/or quantitative evaluation of the nature of the adverse health effects associated with biological, chemical and physical agents that may be present in food. For chemical agents, a dose-response assessment should be performed. For biological or physical agents, a dose-response assessment should be performed if the data are obtainable (MRA). [1]

Hazard identification The identification of biological, chemical and physical agents capable of causing adverse health effects and that may be present in a particular food or group of foods (MRA). [1]

Illness   A condition marked by pronounced deviation from the normal healthy state. [12]

Independent action The mean probability of infection per inoculated microorganism is independent of the number of organisms in the inoculum (disease). [4]

Infection The entry and development of an infectious agent in the body of man or animals (disease). [3]

Infectious pathogens, toxico-infectious pathogens and **toxinogenic pathogens** Three broad classes of foodborne pathogens are differentiated – infectious, toxico-infectious or toxigenic – based on their modes of pathogenicity. Infectious pathogens typically have a three-step process by which they elicit a disease response: ingestion of viable cells, the attachment of these cells to specific locations along the gastro-intestinal tract (or some other mechanisms for avoiding being swept away due to peristalsis), and the invasion of either the epithelium (gastroenteritis) or the body proper (septicaemia). Toxico-infectious agents follow a similar three-step process, except that instead of invading the epithelium or body, they remain in the gastro-intestinal tract, where they either produce or release toxins that affect sites of the epithelium and/or within the body. Toxinogenic bacteria are differentiated on the basis that they cause disease by producing toxins in foods prior to its ingestion. [16]

Inherent randomness Random perturbations that are irreducible in principle, such as Heisenberg's Uncertainty Principle.

Inputs That which is put in or taken in, or which is operated on or utilized by any process or system (either material or abstract), e.g. the information that is put into a model.

Inter-individual variability see *Variability.*

Intra-individual variability see *Variability.*

Likelihood The probability of the observed data for various values of the unknown model parameters (statistics). [3]

Markov chain Monte Carlo A general method of sampling arbitrary highly-dimensional probability distributions by taking a random walk through configuration space. One changes the state of the system randomly according to a fixed transition rule, thus generating a random walk through state space, s0,s1,s2, ... The definition of a Markov process is that the next step is chosen from a probability distribution that depends only on the present position. This makes it very easy to describe mathematically. The process is often called the drunkard's walk (statistics). [9]

Model A set of constraints restricting the possible joint values of several quantities. A hypothesis or system of belief regarding how a system works or responds to changes in its inputs. The purpose of a model is to represent a particular system of interest as accurately and precisely as necessary with respect to particular decision objectives.

Model boundaries Designated areas of competence of the model, including time, space, pathogens, pathways and exposed populations, and acceptable ranges of values for each input and jointly among all inputs for which the model meets data quality objectives.

Model detail Level of simplicity or detail associated with the functional relationships assumed in the model compared to the actual but unknown relationships in the system being modelled.

Model structure A set of assumptions and inference options upon which a model is based, including underlying theory as well as specific functional relationships.

Model uncertainty Bias or imprecision associated with compromises made or lack of adequate knowledge in specifying the structure and calibration (parameter estimation) of a model.

Outbreak (foodborne) An incident in which two or more persons experience a similar illness after ingestion of the same food, or after ingestion of water from the same source, and where epidemiological evidence implicates the food or water as the source of the illness.

Parameter A quantity used to calibrate or specify a model, such as parameters of a probability model (e.g. mean and standard deviation for a normal distribution). Parameter values are often selected by fitting a model to a calibration data set.

Poisson distribution Poisson distributions model (some) discrete random variables (i.e. variables that may take on only a countable number of distinct values, such as 0, 1, 2, 3, 4, ...). Typically, a Poisson random variable is a count of the number of events that occur in a certain time interval or spatial area (statistics). [6]

Precision A measure of the reproducibility of the predictions of a model or repeated measurements, usually in terms of the standard deviation or other measures of variation among such predictions or measurements.

Probability Defined depending on philosophical perspective:

1. The frequency with which we obtain samples within a specified range or for a specified category (e.g. the probability that an average individual with a particular mean dose will develop an illness).

2. Degree of belief regarding the likelihood of a particular range or category.

Probabilistic analysis Analysis in which distributions are assigned to represent variability or uncertainty in quantities. The form of the output of a probabilistic analysis is likewise a distribution.

Probability distribution A function that for each possible value of a *discrete* random variable takes on the probability of that value occurring, or a curve which specifies by means of the area under the curve over an interval the probability that a *continuous* random variable falls within the interval (the probability density function).

Qualitative risk assessment A risk assessment based on data that, while forming an adequate basis for numerical risk estimations, nonetheless, when conditioned by prior expert knowledge and identification of attendant uncertainties, permits risk ranking or separation into descriptive categories of risk. [10]

Quantitative risk assessment A risk assessment that provides numerical expressions of risk and indication of the attendant uncertainties. [10]

Quorum sensing Quorum sensing is a form of communication between bacteria based on the use of signalling molecules that allows bacteria to coordinate their behaviour. The accumulation of signalling molecules in the environment enables a single cell to sense the number of bacteria (cell density). Behavioural responses include adaptation to availability of nutrients, defence against other microorganisms that may compete for the same nutrients, and the avoidance of toxic compounds potentially dangerous for the bacteria. For example, it is very important for pathogenic bacteria during infection of a host (e.g. humans, other animals or plants) to coordinate their virulence in order to escape the immune response of the host in order to be able to establish a successful infection.

Random error Unexplainable but characterizable variations in repeated measurements of a fixed true value resulting from processes that are random or statistically independent of each other, such as imperfections in measurement techniques. Some random errors could be reduced by developing improved techniques.

Refined method This method is intended to provide accurate exposure and risk using appropriately rigorous and scientifically credible methods. The purpose of such methods, models or techniques is to produce an accurate and precise estimate of exposure or risk, or both, consistent with data quality objectives or best practice, or both.

Representativeness The property of a sample (set of observations) that they are characteristic of the system from which they are a sample or which they are intended to represent, and thus appropriate to use as the basis for making inferences. A representative sample is one that is free of unacceptably large bias with respect to a particular data quality objective.

Risk A function of the probability of an adverse health effect and the severity of that effect, consequential to a hazard(s) in food. [1]

Risk analysis A process consisting of three components: *risk assessment, risk management* and *risk communication*. [1]

Risk assessment A scientifically-based process consisting of the following steps : (i) *hazard identification*, (ii) *hazard characterization*, (iii) *exposure assessment*, and (iv) *risk characterization*. [1]

Risk characterization The qualitative and/or quantitative estimation, including attendant uncertainties, of the probability of occurrence and severity of known or potential adverse health effects in a given population based on *hazard identification, hazard characterization* and *exposure assessment* (MRA). [1]

Risk communication The interactive exchange of information and opinions throughout the risk analysis process, concerning hazards and risks, risk-related factors and risk perceptions, among risk assessors, risk managers, consumers, industry, academic community and other interested parties, including the explanation of *risk assessment* findings and the basis for risk management decisions. [1]

Risk estimate The output of risk characterization. [10]

Risk management The process, distinct from *risk assessment*, of weighting policy alternatives, in consultation with all interested parties, considering *risk assessment* and other factors relevant for the health protection of consumers and the promotion of fair trade practices, and, if needed, selecting appropriate prevention and control options. [1]

Sampling distribution A probability distribution for a statistic.

Scenario A construct characterizing the likely pathway affecting the safety of the food product. This may include consideration of processing, inspection, storage, distribution and consumer practices. Probability and severity values are applied to each scenario. [17]

Sensitivity analysis A method used to examine the behaviour of a model by measuring the variation in its outputs resulting from changes in its inputs. [10]

Screening method This method is intended to provide conservative overestimates of exposure and risk using relatively simple and quick calculation methods and with relatively low data input requirements. The purpose of such methods, models or techniques is to eliminate the need for further, more detailed modelling for scenarios that do not cause or contribute to high enough levels of exposure or risk to be of potential concern. If a screening method indicates that levels of exposure or risk are low, then there should be high confidence that actual exposures or risk levels are low. Conversely, if a screening method indicates that estimated exposure or risk levels are high, then a more refined method should be applied since the screening method is intentionally biased. See *Refined method*.

Statistic A function of a random sample of data (e.g. mean, standard deviation, distribution parameters).

Stochastic uncertainty Also referred to as *random error,* q.v.

Stochastic variability Sources of heterogeneity of values associated with members of a population that are a fundamental property of a natural system and that in practical terms cannot be modified, stratified or reduced by any intervention. For example, variation in human susceptibility to illness for a given dose for which there is no predictive capability to distinguish the response of a specific individual from that of another. Stochastic variability contributes to overall variability for measures of individual risk and for population risk.

Subjective probability distribution A probability distribution that represents an individual's or group's belief about the range and likelihood of values for a quantity, based upon that person's or group's *expert judgement,* q.v.

Surrogate data Substitute data or measurements on one quantity used to estimate analogous or corresponding values for another quantity.

Systematic error see *bias*.

Transparent Characteristics of a process where the rationale, the logic of development, constraints, assumptions, value judgements, limitations and uncertainties of the expressed determination are fully and systematically stated, documented and accessible for review. [10]

Threshold Dose of a substance or exposure concentration below which a stated effect is not observed or expected to occur (disease). [5]

Toxigenic pathogens – see *infectious pathogens*.

Toxico-infectious pathogens – see *infectious pathogens*.

Uncertainty Lack of knowledge regarding the true value of a quantity, such as a specific characteristic (e.g. mean, variance) of a distribution for variability, or regarding the appropriate and adequate inference options to use to structure a model or scenario. These are also referred to as *model uncertainty* and *scenario uncertainty*. Lack of knowledge uncertainty can be reduced by obtaining more information through research and data collection, such as through research on mechanisms, larger sample sizes or more representative samples.

Validation Comparison of predictions of a model to independently estimated or observed values of the quantity or quantities being predicted, and quantification of biases in mean prediction and precision of predictions.

Variability Observed differences attributable to true heterogeneity or diversity in a population or exposure parameter. Variability implies real differences among members of that population. For example, different individuals have different intakes and susceptibility. Differences over time for a given individual are referred to as **intra-individual variability.** Differences over members of a population at a given time are referred to as **inter-individual variability**. Variability in microbial risk assessment cannot be reduced but only more precisely characterized.

— § —

Sources of definitions

[1] Codex Alimentarius Commission. 2001 Procedural manual. Twelfth edition. Rome, Food and Agriculture Organization of the United Nations and World Health Organization.

[2] WHO. 1994. Assessing human health risks of chemicals: derivation of guidance values for health-based exposure limits. *Environmental Health Criteria*, No. 170.

[3] Last, J.M. (ed). 1995. *A dictionary of epidemiology*. 3rd ed. New York, NY: Oxford University Press.

[4] Meynell, G.G., & Stocker, B.A.D. 1957. Some hypotheses on the aetiology of fatal infections in partially resistant hosts and their application to mice challenged with *Salmonella paratyphi-B* or *Salmonella typhimurium* by intraperitoneal injection. *Journal of General Microbiology*, 16: 38–58.

[5] WHO. 1999. Risk Assessment Terminology: methodological considerations and provisional results. Report on a WHO experiment. *Terminology Standardization and Harmonization*, vol. 2, nos. 1–4.

[6] http://www.stats.gla.ac.uk/steps/glossary/sampling.html

[7] http://linkage.rockefeller.edu/wli/glossary/stat.html

[8] http://inside.uidaho.edu/tutorial/gis/engine.asp?term=goodness-of-fit

[9] http://www.mcc.uiuc.edu/SummerSchool/David%20Ceperley/dmc_lec3.htm

[10] Codex Alimentarius Commission. 1999. Principles and guidelines for the conduct of microbiological risk assessment. Doc. No. CAC/GL-30.

[11] ILSI [International Life Science Institute]. 2000. Revised framework for microbial risk assessment. ILSI, Washington.

[12] *Dorland's illustrated medical dictionary*. Twenty-sixth edition. 1981. Philadelphia. W.B. Saunders Company.

[13] Benenson, A.S. (ed). 1995. *Control of communicable diseases manual*. Sixth edition. Washington DC: American Public Health Association.

[14] Anonymous. 2003. Risk assessment of food borne bacterial pathogens: quantitative methodology relevant for human exposure assessment. Brussels, European Commission, Health & Consumer Protection Directorate-General.

[15]. Gregg, M., Dicker, R.C., & Goodman, R.A. (eds). 1996. *Field epidemiology*. New York, NY: Oxford Press.

[16] Buchanan, R.L., Smith, J.L., & Long, W. 2000. Microbial risk assessment: dose-response relations and risk characterization. *International Journal of Food Microbiology,* 58: 159–172.

[17] FAO/WHO. 1995. Application of risk analysis to food standards issues. Report of a joint FAO/WHO expert consultation, Geneva, Switzerland, 13–17 March 1995. WHO, Geneva.

[18] Burnham, K.P., & Anderson, D.R. 1998. *Model selection and inference.* Springer.